Transformation Groups for Beginners

STUDENT MATHEMATICAL LIBRARY
Volume 25

Transformation Groups for Beginners

S. V. Duzhin
B. D. Chebotarevsky

This work was originally published in Russian by Вышэйшая Школа, Minsk, under the title От Орнаментов до Дифференциальных Уравнений in 1998. The present translation was created under license for the American Mathematical Society and is published by permission.

Translated from the Russian by S. V. Duzhin.

2000 *Mathematics Subject Classification.* Primary 22-01, 54H15.

For additional information and updates on this book, visit
www.ams.org/bookpages/stml-25

Library of Congress Cataloging-in-Publication Data

Duzhin, S. V. $q (Sergei Vasil'evich), 1956–
[Ot ornamentov do differentsial'nykh uravnenii. English]
Transformation groups for beginners / S. V. Duzhin, B. D. Chebotarevsky; [translated from the Russian by S. V. Duzhin].
p. cm. – (Student mathematical library, ISSN 1520-9121; v. 25)
Includes index.
Romanized record.
ISBN 0-8218-3643-9 (acid-free paper)
1. Transformation groups–Popular works. 2. Algebraic topology–Popular works. I. Chebotarevskii, B. D. $q (Boris Dmitrievich) II. Title. III. Series.
QA385.D8913 2004
512′.55–dc22   2004049676

∞ The paper used in this book is acid-free and falls within the guidelines established to ensure permanence and durability.
Visit the AMS home page at http://www.ams.org/

10 9 8 7 6 5 4 3 2 1   09 08 07 06 05 04

Contents

Preface

The first, Russian, version of this book was written in 1983-1986 by B. D. Chebotarevsky and myself and published in 1988 by "Vysheishaya Shkola" (Minsk) under the title "From ornaments to differential equations". The pictures were drawn by Vladimir Tsesler.

Years went by, and I kept receiving positive opinions about the book from both acquaintances and strangers. In 1996 I decided to translate the book into English. In the course of doing so, I tried to make the book more consistent and self-contained. I deleted some unimportant fragments and added several new sections. Also, I corrected many mistakes (I can only hope I did not introduce new ones).

The translation was finished by the year 2000, and in that year the English text was further translated into Japanese and published by Springer Verlag Tokyo under the title "Henkangun Nyūmon" ("Introduction to Transformation Groups").

The book is intended for college and graduate students. Its aim is to introduce the concept of a transformation group, using examples from different areas of mathematics. In particular, the book includes an elementary exposition of the basic ideas of Sophus Lie related to symmetry analysis of differential equations, which have not yet appeared in the popular literature.

The book contains many exercises with hints and solutions, which will help a diligent reader to master the material.

The present version, updated in 2002, incorporates some new changes, including the correction of errors and misprints kindly indicated by the Japanese translators S. Yukita (Hosei University, Tokyo) and M. Nagura (Yokohama National University).

S. Duzhin
St. Petersburg
September 1, 2002

Introduction

Probably, the one most famous book in the whole history of mathematics is Euclid's "Elements". In Europe it was used as a standard textbook of geometry in all schools for about 2000 years.

One of the first theorems in the "Elements" is the following Proposition I.5, of which we quote only the first half.

Theorem 1 (Euclid). *In isosceles triangles the angles at the base are equal to one another.*

Proof. Every high school student knows the standard modern proof of this proposition. It is very short.

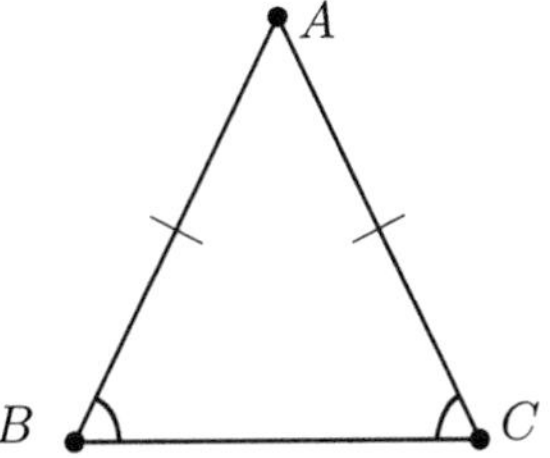

Figure 1. An isosceles triangle

STANDARD PROOF. Let ABC be the given isosceles triangle (Figure 1). Since $AB = AC$, there exists a plane movement (reflection) that takes A to A, B to C and C to B. Under this movement, $\angle ABC$ goes into $\angle ACB$; therefore, these two angles are equal. $\square$

It seems that there is nothing interesting about this theorem. However, wait a little and look at Euclid's original proof (Figure 2).

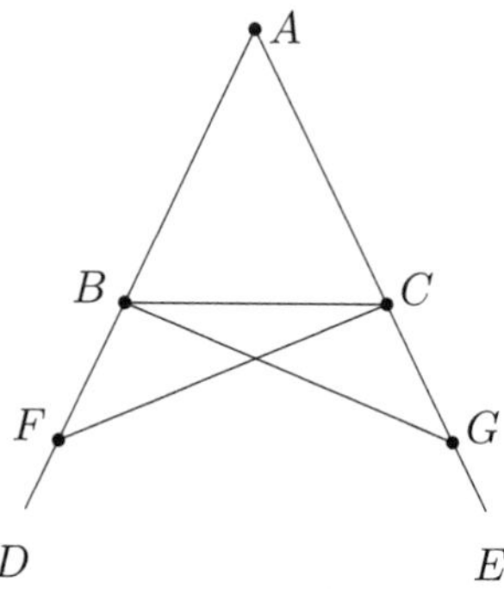

Figure 2. Euclid's proof

EUCLID'S ORIGINAL PROOF. On the prolongations AD and AE of the sides AB and AC, choose points F and G such that $AF = AG$. Then $\triangle ABG = \triangle ACF$; hence $\angle ABG = \angle ACF$. Also $\triangle CBG = \triangle BCF$; hence $\angle CBG = \angle BCF$. Therefore $\angle ABC = \angle ABG - \angle CBG = \angle ACF - \angle BCF = \angle ACB$. □

In mediaeval England, Proposition I.5 was known under the name of *pons asinorum* (asses' bridge). In fact, the part of Figure 2 formed by the points F, B, C, G and the segments that join them really resembles a bridge. Poor students who could not master Euclid's proof were compared to asses that could not surmount this bridge.

From a modern viewpoint Euclid's argument looks cumbersome and weird. Indeed, why did he ever need these auxiliary triangles ABG and ACF? Why was he not happy just with the triangle ABC itself? The reason is that Euclid just could not use *movements* in geometry: this was forbidden by his philosophy, stating that "mathematical objects are alien to motion".

This example shows that the use of movements can elucidate geometrical facts and greatly facilitate their proof. But movements are

Figure 3. Asses's Bridge

important not only when they are studied separately. It is very interesting to study the *social behaviour* of movements, i.e. the structure of *sets of interrelated movements* (or more general transformations). In this area, the most important notion is that of a *transformation group*.

The theory of groups, as a mathematical theory, appeared not so long ago, only in the nineteenth century. However, examples of objects that are directly related to transformation groups had been created back in ancient civilizations, both oriental and occidental. This refers to the art of ornament, called "the oldest aspect of higher mathematics expressed in an implicit form" by the famous twentieth century mathematician Hermann Weyl.

Figure 4 shows two examples of ornaments found on the walls of the mediaeval Alhambra Palace in Spain.

Both patterns are highly symmetric in the sense that they are preserved by many plane movements. In fact, the symmetry properties of Figure 4a are very close to those of Figure 4b: each ornament has an infinite number of translations, rotations by 90° and 180°, reflections and glide reflections. However, they are not identical. The difference between them is in the way these movements are related to

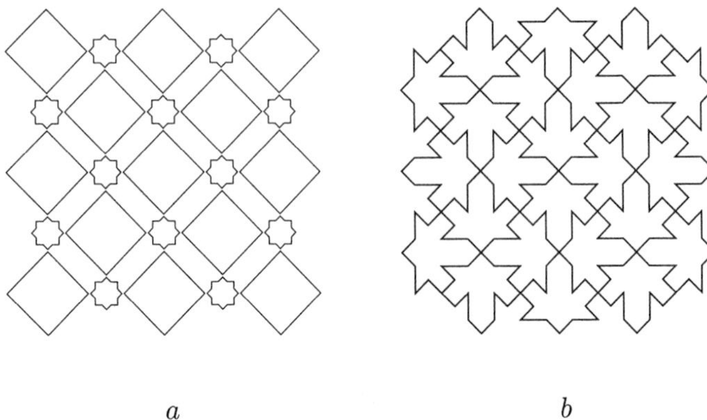

a *b*

Figure 4. Two ornaments from Alhambra

each other for each of the two patterns. The exact meaning of these words can only be explained in terms of group theory, which says that the symmetry groups of Figures 4a and 4b are not isomorphic (this is the contents of Exercise 129, at the end of Chaper 5).

The problem of determining and classifying *all* the possible types of wall pattern symmetry was solved in the late nineteenth century independently by the Russian scientist E. S. Fedorov and the German scientist A. Schoenflies. It turned out that there are exactly 17 different types of plane crystallographic groups (see the table at the end of Chapter 5).

Of course, the significance of group theory goes far beyond the classification of plane ornaments. In fact, it is one of the key notions in the whole of mathematics, widely used in algebra, geometry, topology, calculus, mechanics, etc.

This book provides an elementary introduction into the theory of groups. We begin with some examples from elementary Euclidean geometry, where plane movements play an important role and the ideas of group theory naturally arise. Then we explicitly introduce the notion of a transformation group and the more general notion of an abstract group, and discuss the algebraic aspects of group theory and its applications in number theory. After that we pass to group actions, orbits, invariants, and some classification problems, and finally go as far as the application of continuous groups to the solution of differential equations. Our primary aim is to show how the notion of group works in different areas of mathematics, thus demonstrating that mathematics is a unified science.

The book is intended for people with the beginning of a basic college mathematical education, including the knowledge of elementary algebra, geometry and calculus.

You will find many problems given with detailed solutions, and many exercises, supplied with hints and answers at the end of the book. It goes without saying that the reader who wants to really understand what's going on must try to solve as many problems as possible.

Chapter 1

Algebra of Points

In this chapter we will introduce algebraic operations, addition and multiplication, in the set of points in the plane. This will allow us to apply algebra to geometry and geometry to algebra.

1. Checkered plane

Consider a plane with a regular square grid, i.e. two sets of parallel equal-distanced lines, perpendicular to each other. We will be interested in the polygons with all vertices at nodes of the grid, like the isosceles triangle or the square shown in Figure 1.

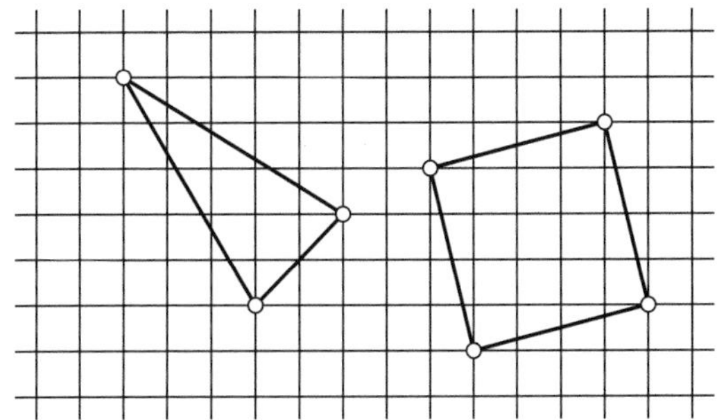

Figure 1. Polygons in the checkered plane

Problem 1. *Prove that a regular polygon different from a square cannot have all its vertices at nodes of a square grid.*

Solution. Suppose, on the contrary, that such a polygon $A_1A_2 \dots A_n$ exists. Let O be its centre. For every triple of consecutive vertices $A_{k-1}A_kA_{k+1}$ find a point B_k which is the fourth vertex of the parallelogram $A_{k-1}A_kA_{k+1}B_k$. The whole construction of Figure 2 goes into itself under the reflection with axis OA_k and under the rotation through $360/n$ degrees around the point O. Therefore every point B_k lies on the corresponding line OA_k, and $B_1B_2 \dots B_k$ is a regular polygon. If $n > 6$, then this polygon is smaller than the initial one. Indeed, in this case the angle $\alpha = \frac{n-2}{n}180°$ is greater than the angle $\beta = \frac{2}{n}360°$; hence the point B_k belongs to the segment OA_k. It is a crucial observation that all the points B_1, B_2,..., B_n lie again at nodes of the square grid.

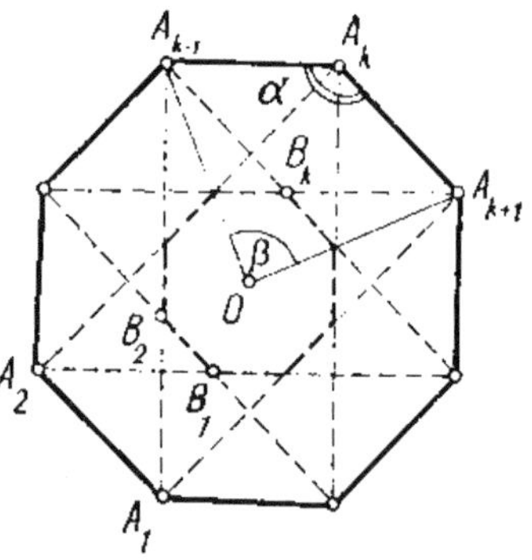

Figure 2. Regular polygon

Repeating the same procedure for the polygon $B_1B_2 \dots B_n$ instead of $A_1A_2 \dots A_n$, we will arrive at a third polygon $C_1C_2 \dots C_n$ whose vertices have the following properties:

- they coincide with nodes of the grid, and
- C_k belongs to the segment OA_k and lies closer to O than B_k.

Since there are only finitely many integer points on the segment OA_k, after several iterations of this procedure we will arrive at a contradiction.

The same argument remains valid also in the case of a regular pentagon, the only difference being that now the point B_k lies on the line passing through O and A_k outside of the segment OA_k.

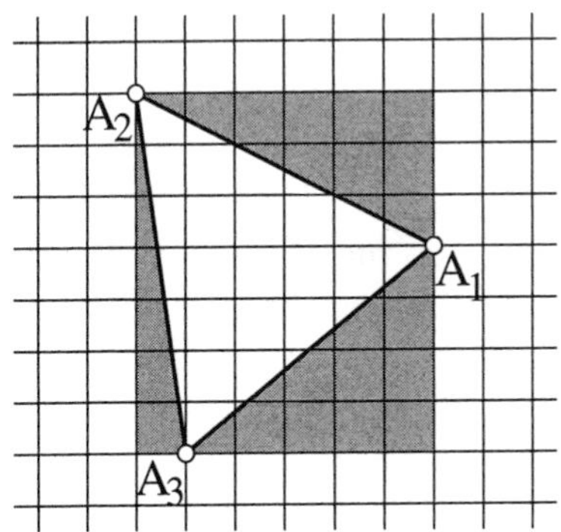

Figure 3. Is there such an equilateral triangle?

If $n = 3$ or 6, the argument fails (why?), and we will give a different proof of our assertion. Note first of all that three vertices of a regular hexagon form an equilateral triangle; thus it is sufficient only to consider the case $n = 3$. Suppose that an equilateral triangle has all its vertices at nodes of the checkered plane (Figure 3). Then, by Pythagoras' theorem, the square of the side of this triangle must be an integer (we assume that the grid is 1 by 1); hence its area $S = a^2\sqrt{3}/4$ is an irrational number. On the other hand, the triangle $A_1A_2A_3$ can be obtained from a rectangle with integer sides by removing three right triangles as shown in Figure 3; thus its area must be rational — in fact, either m or $m + \frac{1}{2}$, where m is a whole number.

Exercise 1. Suppose that the sides of the squares making the grid are 1. Is there a right triangle with all vertices at nodes such that all its sides have integer lengths and no side is parallel to the lines of the grid?

2. Point addition

Our solution of Problem 1 was based on the following nice property of the integer grid: if three vertices of a parallelogram are at nodes, so is the fourth. The usual mathematical wording for this phenomenon is: *the set of all nodes is closed with respect to the operation under study*. We will now give an exact definition of this operation.

Given three points in the plane, say M, N and P, there are three different ways to add one more point so that the triangle MNP becomes a parallelogram. One way is to connect P with the midpoint K of MN and choose the point L on the line PK which is symmetric to P with respect to K (Figure 4).

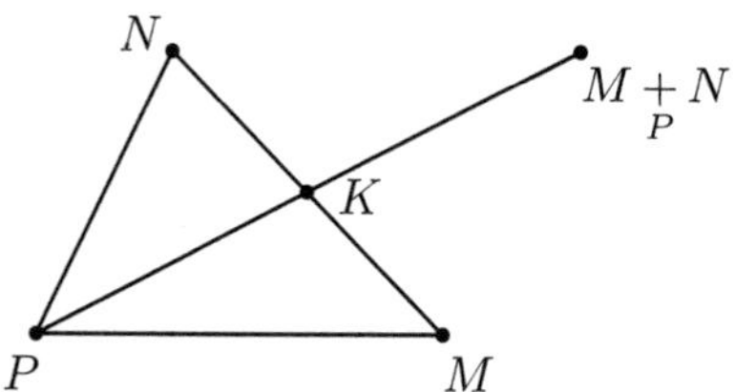

Figure 4. Point addition

Definition 1. We will call the point L thus constructed *the sum of the points M and N over the pole P*, and we will write $L = M \underset{P}{+} N$, which should be read aloud as "M plus N over P". When the pole is fixed, we may omit it from the notation and simply write $L = M + N$.

This definition holds for an arbitrary triple of points in the plane. If M, N and P belong to one straight line, then the parallelogram $MPNL$ degenerates into a line segment. If all of them coincide, then it degenerates even more and becomes a point.

Now we can give an exact statement for the property of the integer grid that was used in Problem 1: *the sum of any two nodes of the grid over any other node is always a node.*

Let us now forget about the grid and study the properties of addition for arbitrary points.

Exercise 2. Given two triangles ABC and DEF and a point P, denote by Φ the set of all points $M \underset{P}{+} N$ where M is an interior point of $\triangle ABC$ and N an interior point of $\triangle DEF$.

a) Prove that Φ is a polygon. How many sides may it have?

b) Prove that its perimeter is the sum of the perimeters of the two given triangles.

Point addition is closely related to vector addition: $L = M \underset{P}{+} N$ is equivalent to $\overrightarrow{PL} = \overrightarrow{PM} + \overrightarrow{PN}$, and enjoys similar properties:

1° The associative law

$$(A \underset{P}{+} B) \underset{P}{+} C = A \underset{P}{+} (B \underset{P}{+} C)$$

holds for any arbitrary points A, B, C over any pole P.

2° We always have

$$P \underset{P}{+} A = A,$$

i.e., the point P behaves as a neutral element with respect to the operation $\underset{P}{+}$.

3° Over a given pole P, every point A has an *opposite* point, i.e., a point A' such that

$$A \underset{P}{+} A' = P.$$

In fact, one can simply take the point A' which is symmetric to A with respect to P.

4° The commutative law

$$A \underset{P}{+} B = B \underset{P}{+} A$$

holds for any three arbitrary points.

The order in which these 4 items appear in our list is not accidental — in fact, more fundamental rules come first. You will understand this better when you get to Chapter 4.

Rules $2^\circ - 4^\circ$ are obvious and do not require any proof. To check rule 1°, we first construct the points $M = A \underset{P}{+} B$ and $N = B \underset{P}{+} C$ (see Figure 5). The segments AM and CN are both equal and parallel to the segment PB. Hence the midpoints of MC and AN coincide, which, by the definition of point addition, ensures that $M \underset{P}{+} C = A \underset{P}{+} N$.

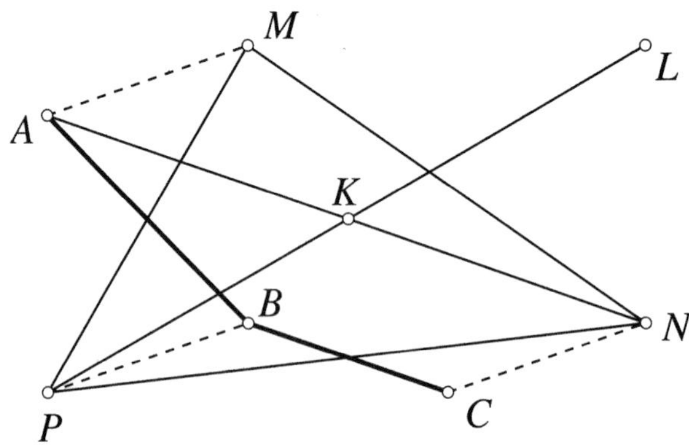

Figure 5. Associativity of point addition

Using property 3°, we can define the *difference* of two points over a given pole: $B \underset{P}{-} A = B \underset{P}{+} A'$, where $A' = \underset{P}{-} A$ is the point opposite to A. The point $B \underset{P}{-} A$ is the unique solution to the equation $A \underset{P}{+} X = B$.

If all operations are carried out over the same pole, then addition and subtraction of points satisfies the same rules as the usual operations on numbers, for example, $A - (B - C + D) = A - B + C - D$.

Problem 2. *Find the sum* $A \underset{M}{+} B \underset{M}{+} C$, *where* M *is the intersection point of the medians in a triangle* ABC.

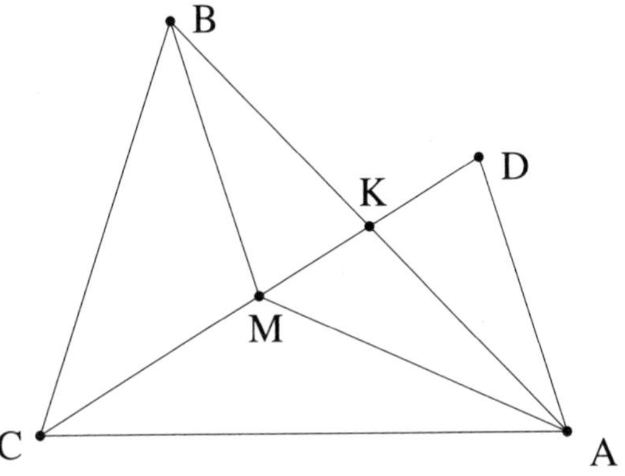

Figure 6. Sum of vertices of a triangle

Solution. Recall that each median is divided by their common intersection point M in the ratio $2:1$; therefore, in Figure 6, we have $CM = 2MK$. The point $D = A \underset{M}{+} B$ lies on the prolongation of the median CK, and $DK = KM = \frac{1}{2}MC$. Therefore, $DM = MC$ and $D \underset{M}{+} C = M$.

It is interesting to observe that the intersection point of the medians is the only point which satisfies $A \underset{M}{+} B \underset{M}{+} C = M$. To prove this, let us first derive the rules for passing from one pole to another in the formulas involving point addition:

$$A \underset{Q}{+} B = A \underset{P}{+} B \underset{P}{-} Q, \tag{1}$$

$$A \underset{Q}{-} B = A \underset{P}{-} B \underset{P}{+} Q. \tag{2}$$

The first equality can be rewritten as $(A \underset{Q}{+} B) \underset{P}{+} Q = A \underset{P}{+} B$, and its validity is easily seen from Figure 7. To prove the second one, we will check that the point $A \underset{P}{-} B \underset{P}{+} Q$ is a solution to the equation $B \underset{Q}{+} X = A$. Indeed, using the formula we have just proved, we get

$$B \underset{Q}{+} (A \underset{P}{-} B \underset{P}{+} Q) = B \underset{P}{+} (A \underset{P}{-} B \underset{P}{+} Q) \underset{P}{-} Q = A.$$

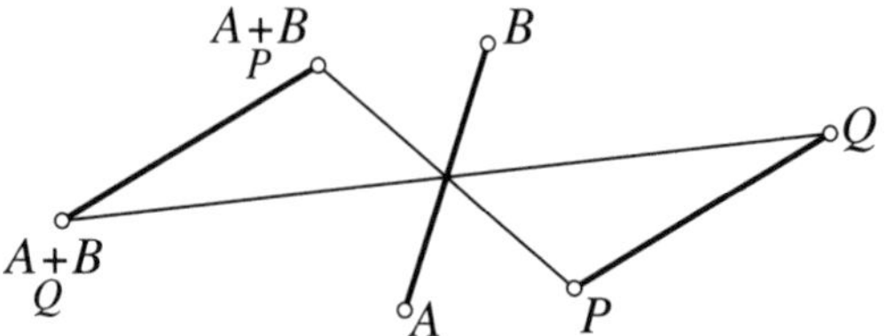

Figure 7. Change of base point

Note that the point P does not appear in the left-hand sides of (1) and (2); hence the right-hand sides do not depend on its choice. This observation is generalized in the following exercise.

Exercise 3. Investigate the conditions under which the expression

$$A_1 \underset{P}{+} A_2 \underset{P}{+} \cdots \underset{P}{+} A_k \underset{P}{-} B_1 \underset{P}{-} B_2 \underset{P}{-} \cdots \underset{P}{-} B_l$$

does not depend on the choice of the pole P.

Continuing the discussion of Problem 2, suppose that a point N has the same property as the median intersection point M, i.e. $A \underset{N}{+} B \underset{N}{+} C = N$. We can subtract the pole without violating the equation; hence $A \underset{N}{+} B \underset{N}{+} C \underset{N}{-} N \underset{N}{-} N = N$. A reader who has done Exercise 3 knows that the left-hand side of this relation does not depend on the choice of N. In particular, substituting M in place of N, we get $A \underset{M}{+} B \underset{M}{+} C \underset{M}{-} N \underset{M}{-} N = N$; therefore $M \underset{M}{-} N \underset{M}{-} N = N$, $N \underset{M}{+} N \underset{M}{+} N = N$ and finally $N = M$. This means that the median intersection point M is the unique point with the property proved in Problem 2.

Exercise 4. Prove that $A \underset{O}{+} B \underset{O}{+} C = H$, where O is the centre of the circle circumscribed around the triangle ABC and H is the intersection point of its three altitudes.

3. Multiplying points by numbers

Over a given pole P, a point A can be *multiplied* by a real number α yielding a new point $B = \alpha_P A$.

Definition 2. The *product of a point A by a real number α over the pole P* is the point B that lies on the line PA at the distance $|\alpha||PA|$ from the pole P and on the same side of P as A, if $\alpha > 0$, or on the other side, if $\alpha < 0$.

Figure 8. Multiplication of points by numbers

In other words, this operation means that you stretch the vector $\overrightarrow{PA}$, keeping its initial point P fixed, as if with a pin: $\overrightarrow{PB} = \alpha\overrightarrow{PA}$.

In particular,

(1) any point multiplied by zero over P gives P, and

(2) P multiplied by any real number gives P.

It is easy to see that multiplication of points by numbers has these properties:

5° $1_P A = A$.

6° $\alpha_P(\beta_P A) = (\alpha\beta)_P A$.

7° $(\alpha + \beta)_P A = \alpha_P A \underset{P}{+} \beta_P A$.

8° $\alpha_P(A \underset{P}{+} B) = \alpha_P A \underset{P}{+} \alpha_P B$.

To multiply a point by a *natural* number n is the same thing as to add up n equal points: $n_P A = A \underset{P}{+} A \underset{P}{+} \ldots \underset{P}{+} A$ (A repeated n times). Using this fact, you can check that the point $\frac{1}{2}{}_P A$ is the (unique) solution to the equation $X \underset{P}{+} X = A$.

Consider a *linear combination over the pole P*, i.e. the sum of several points with arbitrary coefficients

$$\alpha_P A \underset{P}{+} \beta_P B + \ldots \underset{P}{+} \omega_P Z = S. \tag{3}$$

In general, the resulting point S depends on the choice of the pole P. When $\alpha, \beta, \ldots, \omega$ are integer numbers, we have seen in Exercise 3 that there are some occasions when the result does not depend on P. This may also happen in the more general situation when the coefficients are not integer. For example, the point $M = \frac{1}{2}A + \frac{1}{2}B$ is always the middle point of the segment AB, wherever you put the pole.

Exercise 5 (A generalization of Exercise 3). Find a necessary and sufficient condition on the coefficients α, β, ..., ω which guarantees that the linear combination S of (3) does not depend on the choice of the point P.

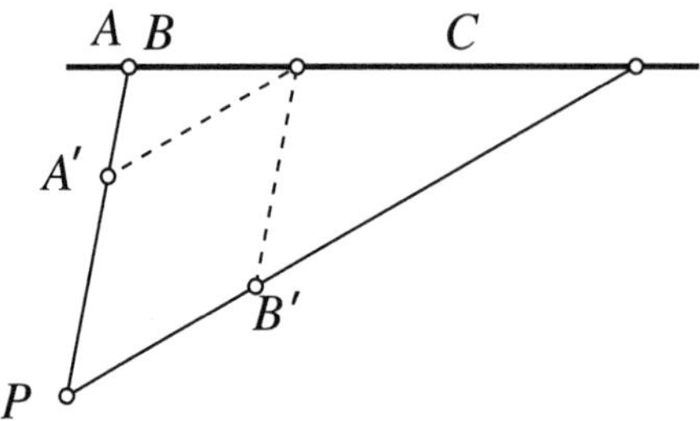

Figure 9. Point of a segment expressed in terms of endpoints

Using point addition and multiplication by numbers, it is possible to express any point of the segment AB in terms of its endpoints. Indeed, suppose that the point C divides the segment AB in the ratio $k : l$ (by definition, this means that $l \cdot \overrightarrow{AC} = k \cdot \overrightarrow{CB}$). Choose an arbitrary point P outside of the line AB; we will use it as the pole in all subsequent operations on points. Through the point C we draw two lines, parallel to PB and PA, which meet PA and PB at points A' and B', respectively (see Figure 9). Then

$$\frac{PA'}{PA} = \frac{BC}{BA} = \frac{l}{k+l},$$
$$\frac{PB'}{PB} = \frac{AC}{AB} = \frac{k}{k+l}.$$

Put $\frac{l}{k+l} = \alpha$, $\frac{k}{k+l} = \beta$. Then $C = A' + B' = \alpha A + \beta B$, and the sum of the two coefficients α and β is 1.

The converse is also true: if α and β are arbitrary nonnegative numbers such that $\alpha + \beta = 1$, then the point $C = \alpha A + \beta B$ belongs to the segment AB. Moreover, if one of the numbers α and β in the formula for C is negative, but the sum of the two is still 1, then the point C lies on the straight line AB, but outside of the segment

AB. By changing Figure 9 appropriately, you can verify that the relations $\alpha = \frac{l}{k+l}$, $\beta = \frac{k}{k+l}$ remain valid, although the ratio $k : l$ is now negative.

Thus, the straight line AB is the set of all points $\alpha A + (1-\alpha)B$, where α is an arbitrary real number, while the segment AB is its subset specified by the restriction $0 \le \alpha \le 1$. Note again that this description does not depend on the choice of the base point (pole).

Exercise 6. Find a similar description of the set of all interior points of a convex polygon with vertices $A_1, A_2, \ldots, A_n$.

After doing Exercise 6, you can go back and tackle Exercise 2 once again, using the new technique.

Exercise 7. A middle line of a quadrilateral is the line joining the midpoints of two opposite sides. Any quadrilateral has two middle lines. Prove that these two lines, as well as the segment joining the midpoints of the two diagonals, meet in one point, and this point divides each of them in half (see Figure 10).

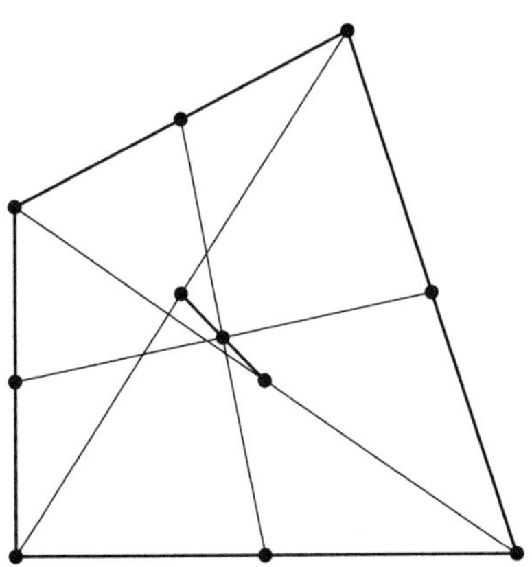

Figure 10. Quadrilateral of exercise 7

4. Centre of gravity

In Problem 2, the median intersection point M of a triangle ABC was described implicitly as the (unique!) solution to the equation $A \underset{M}{+} B \underset{M}{+} C = M$. We can now express M explicitly in terms of A, B

and C. Indeed, multiplying both sides of the equation by $\frac{1}{3}$, we get $\frac{1}{3}{}_M(A \underset{M}{+} B \underset{M}{+} C) = M$. According to the answer of Exercise 5, the left-hand side of this equality does not depend on the choice of the pole; hence we can write

$$M = \frac{1}{3}(A + B + C).$$

In a similar way, the point referred to in Exercise 7 can be expressed in terms of the vertices of the quadrilateral as

$$M = \frac{1}{4}(A + B + C + D).$$

In general, the arithmetic mean of several points is called the *centre of gravity* (or *centre of mass*) of the system consisting of these points: $M = \frac{1}{n}(A_1 + A_2 + \cdots + A_n)$ (over an arbitrary pole). Thus, the centre of gravity of a triangle (or, more exactly, of the set of its vertices) is the median intersection point, while the centre of gravity of the set of vertices of a quadrilateral is the intersection of its two middle lines.

We proceed to some examples where geometric problems related to the centre of gravity are solved using operations on points.

Problem 3. *Suppose that A, B and C are three collinear points, while E and F are arbitrary points in the plane. Prove that the median intersection points of the triangles AEF, BEF, CEF are collinear.*

Solution. Median intersection points are arithmetic means of the vertices:

$$\begin{aligned} \frac{1}{3}(A + E + F) &= K, \\ \frac{1}{3}(B + E + F) &= L, \\ \frac{1}{3}(C + E + F) &= M. \end{aligned}$$

By assumption, point C lies on the line AB, thus $C = \alpha A + (1 - \alpha)B$. Hence

$$\begin{aligned} \alpha K + (1 - \alpha)L &= \frac{\alpha}{3}(A + E + F) \\ &= \frac{1 - \alpha}{3}(B + E + F) = \frac{1}{3}(C + E + F) = M, \end{aligned}$$

which implies that the point M belongs to the line KL.

Exercise 8. Let A, B, C, D, E, F be the middle points of the consecutive edges of a hexagon. Prove that the centres of gravity of the triangles ACE and BDF coincide.

Exercise 9. In a quadrilateral $ABCD$, the point E is the midpoint of the side AB and K the midpoint of the side CD. Prove that the midpoints of the four segments AK, CE, BK and ED form a parallelogram.

Problem 4. *Prove that the middle line of a quadrilateral (see Exercise 7) passes through the intersection point of its diagonals if and only if this quadrilateral is a trapezoid, i.e. has two parallel sides.*

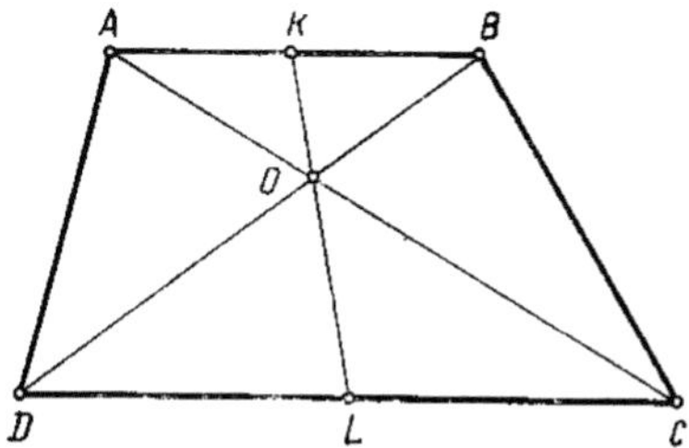

Figure 11. Trapezoid

Solution. We choose the intersection point O of the diagonals as the pole (see Figure 11). Then $C = \alpha A$, $D = \beta B$ for appropriate numbers α and β, and for the middle points K and L we can write $K = \frac{1}{2}(A + B)$, $L = \frac{1}{2}(\alpha A + \beta B)$.

If $AB \parallel CD$, then the triangles OBA and ODC are similar, hence $\alpha = \beta$, $L = \alpha K$ and the points K, L, O are collinear.

Suppose, on the other hand, that we do not know whether AB is parallel to CD, but we do know that K, L and O lie on the same line. Then, using the point operations over the pole O, for a suitable real number γ

we have $L = \gamma K$. Substituting the previous expressions for K and L, we get $\alpha A + \beta B = \gamma A + \gamma B$, or $(\alpha - \gamma)A = (\gamma - \beta)B$. But the points $(\alpha - \gamma)A$ and $(\gamma - \beta)B$ lie on different lines OA and OB, and if they coincide, this must mean that they coincide at O. Thus, $\alpha - \gamma = \gamma - \beta = 0$, $\alpha = \beta$, triangles OAB and ODC are similar, and $AB \parallel CD$.

Exercise 10. Using point addition and multiplication by numbers, find an independent proof of the fact that the medians of a triangle are divided in proportion 2 : 1 by their intersection point (note that we have used this fact before, in Problem 2).

Exercise 11. A line cuts 1/3 of one side of a parallelogram and 1/4 of the adjacent side in such a way that the smaller parts have a common vertex (Figure 12). In what ratio does this line divide the diagonal of the parallelogram?

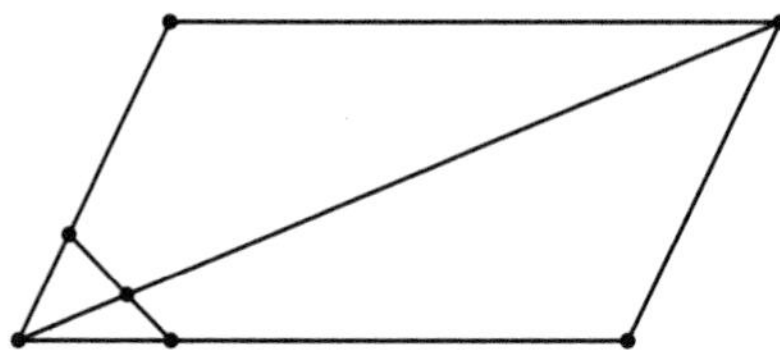

Figure 12. Cutting the diagonal

5. Coordinates

In the discussion of Problem 4, we have used the following important fact: *if two points M, N are not collinear with the pole, then the equality $\alpha M + \beta N = \gamma M + \delta N$ is possible only if $\alpha = \beta$ and $\gamma = \delta$.* In fact, the given equality can be rewritten as $(\alpha - \gamma)M = (\gamma - \delta)N$, which implies $\alpha = \gamma$ and $\beta = \delta$.

Choose a pole P and two points M, N that are not collinear with P. Then any point Z of the plane can be expressed as $Z = xM + yN$ for suitable real numbers x and y (Figure 13).

Definition 3. A *system of affine coordinates* in the plane is an ordered set of three non-collinear points $\{P, M, N\}$. The first point P is

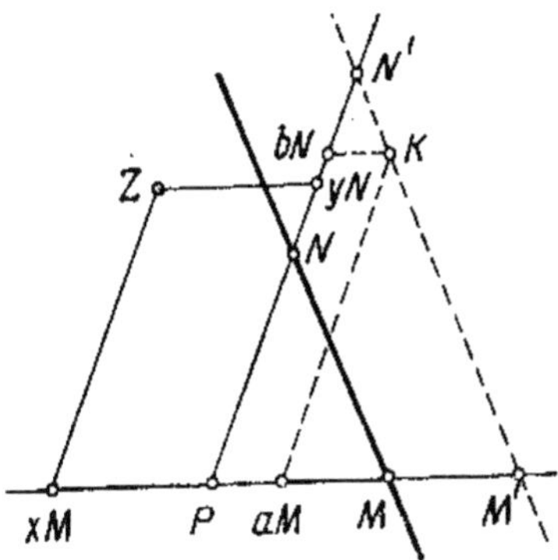

Figure 13. Affine coordinates

referred to as the *pole*, or the *origin*, while the set $\{M, N\}$ is referred to as the *basis* of the given coordinate system. The coordinates of a point Z in the coordinate system $\{P, M, N\}$ are the coefficients $\{x, y\}$ in the expansion $Z = x_P M + y_P N$.

The above argument shows that the coordinates x and y are *uniquely* determined by the point Z. Thus we obtain a one-to-one correspondence between the points of a plane and pairs of real numbers.

If $\angle MNP$ is a right angle and both PM and PN are unit segments, then what we get is the usual Cartesian coordinate system. In general, such coordinates are referred to as *affine coordinates*.

When two points are added, their coordinates add up:

$$(aM + bN) + (cM + dN) = (a + b)M + (c + d)N.$$

When a point is multiplied by a number, its coordinates get multiplied by the same number:

$$c(aM + bN) = (ca)M + (cb)N.$$

The correspondence between points of the plane and pairs of real numbers can be used as a dictionary which serves to translate geometric propositions into the language of algebra and vice versa. Any geometric figure is the set of all points whose coordinates satisfy a certain relation. For example, we know that a point belongs to the line MN if and only if has the expression $xM + yN$ with $x + y = 1$. In this sense, $x + y = 1$ is the *equation of the straight line* MN.

Problem 5. *Find the equation of the straight line which is parallel to* MN *and passes through the given point* K *with coordinates* a, b.

Solution. Let M' and N' be the intersection points of this line with PM and PN, respectively. Since $M'N' \parallel MN$, we have $M' = tM$, $N' = tN$ for an appropriate number t (see Figure 13). Any point Z of the line $M'N'$ is equal to $\alpha M' + \beta N'$, where $\alpha + \beta = 1$, i.e., $Z = \alpha tM + \beta tN$ and $\alpha t + \beta t = t$. Thus, the coordinates $x = \alpha t$, $y = \beta t$ of an arbitrary point $Z \in MN$ satisfy the relation $x + y = t$, where the value of t is yet unknown. To find it, note that the point K lies on the line under study; hence its coordinates a, b satisfy the equation of this line: $a + b = t$ is true. We have found that $t = a + b$, and the answer to the exercise is $x + y = a + b$.

Exercise 12. Write the equation of the straight line that contains a given point $K(a, b)$ and

(a) is parallel to PM,

(b) is parallel to PN,

(c) passes through P.

Problem 6. *Suppose that in a certain triangular region of the plane the laws of optics are such that a ray of light which goes parallel to one side of the triangle and hits the second side, after reflection assumes the direction of the third side of the triangle. Prove that a person standing inside this triangle and directing the beam of his flashlight parallel to one of the sides of the triangle, is in fact shining the light onto his own back.*

Solution. Let one vertex of the triangle, P, be the pole and two others, M and N, be the two basic points of a coordinate system (Figure 14).

Suppose that the person with the flashlight stands at the point $K(a, b)$ and the beam of his flashlight goes

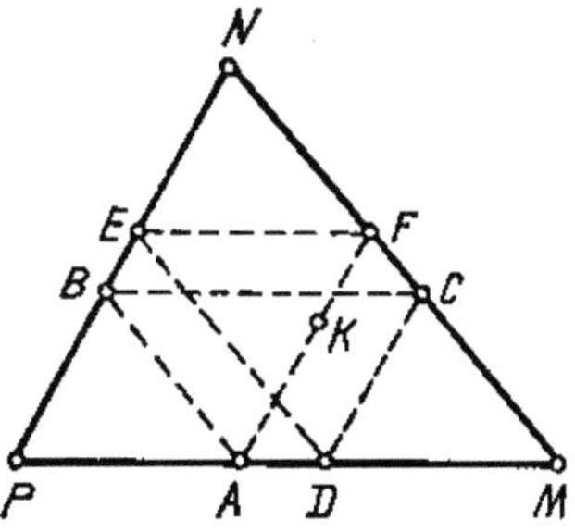

Figure 14. Zigzag inside a triangle

parallel to PN and meets the side PM at the point A. The coordinates of A are $(a, 0)$, because, on one hand, $KA \parallel PN$ and hence the first coordinate of A is equal to the first coordinate of K (see Exercise 12), while on the other hand, point A lies on PM and hence its second coordinate is 0.

The next segment of the beam, AB, is parallel to MN. According to Problem 5, the equation of the line AB is $x + y = a$, because a is the sum of the coordinates of the point A. Since the point B lies on PN, it has $x = 0$; therefore its second coordinate must be equal to a.

Proceeding in the same way, we successively find the coordinates of all points where the beam meets the sides of the triangle: $C(1 - a, a)$, $D(1 - a, 0)$, $E(0, 1 - a)$, $F(a, 1 - a)$. The line FK is parallel to PN, which is why the beam does return to the initial point K — from the opposite direction.

A vigilant reader may have noticed a flaw in the previous argument: in fact, it may happen that the beam returns to the point K before it makes the complete tour of $ABCDEF$ — and hits the flashlighter in a side, not in the back.

Exercise 13. Describe the set of all points K in the triangle MNP for which the trajectory of the flashlight beam consists of only three segments, not six.

Exercise 14. A point K lies inside the triangle ABC. Straight lines AK, BK, CK meet the sides BC, CA, AB at the points D, E, F, respectively (Figure 15). Prove that $KD/AD + KE/BE + KF/CF = 1$.

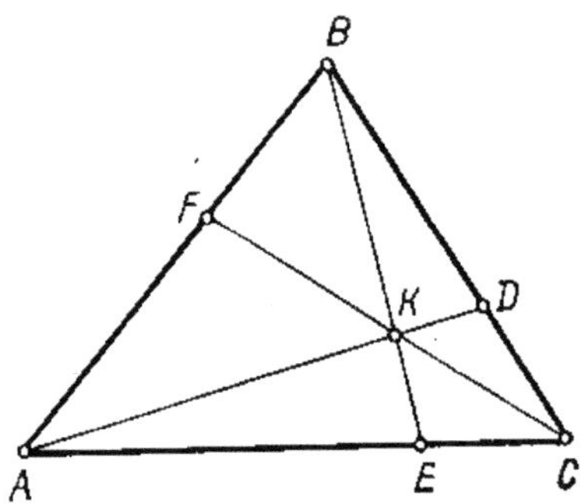

Figure 15. Lines in a triangle meeting at one point

Exercise 15. Given three points D, E, F on the sides of the triangle ABC (Figure 15), prove that the lines AD, BE, CF pass through one point if and only if $AF \cdot BD \cdot CE = FB \cdot DC \cdot EA$ (theorem of Ceva).

6. Point multiplication

We have learned how to multiply a point in the plane by a real number. Now recall that real numbers can be represented as points lying on a line. Let us insert this line into the plane so that its origin (zero point) coincides with the pole P used to define the point addition and multiplication of points by numbers. The unit point of the real line will be E (see Figure 16a).

Our definition of point addition agrees with the usual addition of real numbers in the sense that if the points A and B correspond to numbers a and b, then the sum $A + B$ (over the pole P) corresponds to the number $a + b$ (Figure 16b).

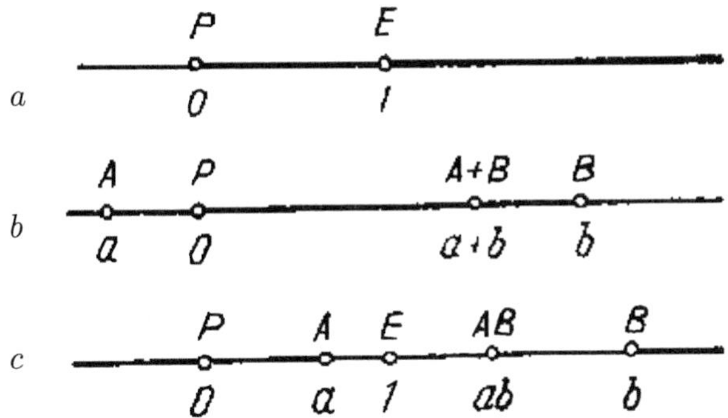

Figure 16. Algebraic operations in the line

Moreover, our multiplication of points by numbers, restricted to the real line, also agrees with the usual product of numbers in the sense that if $A \leftrightarrow a$, $B \leftrightarrow b$, then both points $a_P B$ and $b_P A$ correspond to the number ab. It is natural to call this point the *product* of the two points A and B and denote it by AB.

The next step we want to make is to extend this definition to the entire plane. We want to find a rule to assign a new point AB to any pair of arbitrary points A, B in such a way that this *point multiplication* satisfies the usual rules of multiplication:

9° Associativity

$$(AB)C = A(BC).$$

10° Commutativity

$$AB = BA.$$

11° Distributive law with respect to point addition with the same pole

$$A(B + C) = AB + AC.$$

We also require that the new operation agree with the previously defined multiplication of points by numbers, i.e., that for any point Z in the plane and any point A on the real line that corresponds to the number a we should have $AZ = a_P Z$. In particular, this means that the unit point E of the real line must play the role of the number 1 for all points Z of the plane in the sense that $EZ = 1_P Z = Z$.

It is not immediately clear whether it is possible to introduce such an operation for the points of the plane. We will see, however, that in fact there are many ways to do so, and they come in three essentially different types. But let us first do some exercises.

Problem 7. *Let $EABCDK$ be a regular hexagon with centre at P (recall that E is the unit point), and suppose that $A^2 = B$ for a certain choice of point multiplication. Find all pairwise products of the vertices of the given hexagon.*

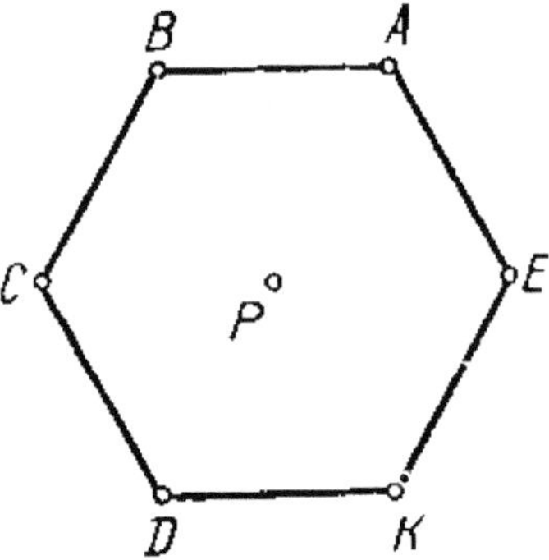

Figure 17. Multiplication of the vertices of a hexagon

Solution. Expand all the vertices over the basis E, A taking P for the pole (Figure 17): $B = A - E$, $C = -E$, $D = -A$, $K = E - A$. We know the products of all pairs consisting of basic points: $E^2 = E$, $EA = A$, $A^2 = B$. Using the distributive law, we can find, for example, that $BK = (A-E)(E-A) = -A^2+2AE-E^2 = -B+2A-E = A$. Other products can be found in similar fashion, yielding the multiplication table:

	E	A	B	C	D	K
E	E	A	B	C	D	K
A	A	B	C	D	K	E
B	B	C	D	K	E	A
C	C	D	K	E	A	B
D	D	K	E	A	B	C
K	K	E	A	B	C	D

Note that the set of 6 vertices of the hexagon turns out to be *closed* under the chosen rule of multiplication, i.e., the product of any two vertices is also a vertex.

Exercise 16. Is the set of vertices of the same hexagon closed under multiplication, if (a) $A^2 = A$; (b) $A^2 = P$? Fill out the corresponding multiplication tables.

Exercise 17. Find the multiplication table for the set of vertices of a regular pentagon $EABCD$ centred at the pole P, if A^2 is known to be equal to B.

After these examples, we can investigate the general case. Suppose that we are given a point multiplication rule that satisfies all the requirements stated above.

Besides the two already chosen points $P = 0$ and $E = 1$, pick an arbitrary point F not on the line PE. Then the pair (E, F) is a basis over P, and, as we saw in the discussion of Problem 7, point multiplication is completely defined, if we only know the square of F.

We have $F^2 = \alpha E + \beta F$ for suitable real numbers α and β. Let us try to find another point G such that the pair (E, G) is also a basis in the plane, but the square G^2 has a simpler expansion over this basis.

Let $G = F - \frac{\beta}{2}E$. Then the lines FG and PE are parallel, so that E and G constitute a basis, and

$$\begin{aligned} G^2 &= \left(F - \frac{\beta}{2}E\right)^2 \\ &= F^2 - \beta EF + \frac{\beta^2}{4}E^2 \\ &= \left(\alpha + \frac{\beta^2}{4}\right)E. \end{aligned}$$

If you look closely at this relation, you will see that multiplication in the basis (E, G) looks simpler than in the initial basis (E, F), because the square of the second basic point is now just E with a certain coefficient — and not the combination of the two points, as before. To further simplify the multiplication rule, we will change G once again, depending on the sign of this coefficient.

(1) $\alpha + \dfrac{\beta^2}{4} = 0$ (cf. Exercise 16b). In this case the product is given by the formulas

$$E^2 = E, \quad EG = G, \quad G^2 = 0;$$
$$(aE + bG)(cE + dG) = acE + (ad + bc)G.$$

(2) $\alpha + \dfrac{\beta^2}{4} > 0$. Denoting $\dfrac{1}{\sqrt{\alpha + \beta^2/4}}G$ by H, in the basis (E, H) we will have the following rules of multiplication:

$$E^2 = E, \quad EH = H, \quad H^2 = E;$$
$$(aE + bH)(cE + dH) = (ac + bd)E + (ad + bc)H.$$

(Try to find such a point H among the vertices of the hexagon in Exercise 16b.)

(3) $\alpha + \dfrac{\beta^2}{4} < 0$. Set $I = \dfrac{1}{\sqrt{|\alpha + \beta^2/4|}}G$. Then

$$E^2 = E, \quad EI = I, \quad I^2 = -E;$$
$$(aE + bI)(cE + dI) = (ac - bd)E + (ad + bc)H.$$

It is easily verified that in each of the three cases our operation satisfies all the laws imposed on multiplication. The next question that naturally appears is whether this multiplication has an inverse operation of *division*, i.e., whether the equation $AZ = B$ can always be solved for Z, provided that $A \neq 0$.

In the first case let us try to divide E by G, i.e., find a point $Z = xE + yG$ such that $GZ = E$. According to the definition, $G(xE+yG) = xG$, which never equals E. Thus, division is in general impossible.

The same is true in the second case, where, as you can check, H is not divisible by $E + H$.

We claim, however, that in the third case division by a non-zero point is always possible. Indeed, let $M = aE + bI$, $N = cE + dI$, where the coefficients c and d do not vanish simultaneously. We want to find the quotient M/N, that is, a point $Z = xE + yI$ such that $NZ = M$, or $(cE + dI)(xE + yI) = aE + bI$. When expanded, this equality becomes equivalent to the system of equations

$$\begin{aligned} cx - dy &= a, \\ dx + cy &= b, \end{aligned}$$

which has a unique solution $x = \dfrac{ac + bd}{c^2 + d^2}$, $y = \dfrac{bc - ad}{c^2 + d^2}$, provided that $c^2 + d^2 \neq 0$.

The result of our investigation can be stated as follows.

Theorem 2. *Multiplication of points in the plane can be introduced in three essentially different ways, depending on the existence of an element X with the property*

(1) $X^2 = 0$,

(2) $X^2 = 1$,

(3) $X^2 = -1$.

Only in case (3) is division by non-zero elements always possible.

Speaking more formally, there exist three different two-dimensional algebras over the field of real numbers, and only one of them (case 3) is an algebra with division.

Note that the actual geometric meaning of multiplication, say, in case (3), depends on the mutual position of points E and I with respect to the origin P. For example, let $PEAB$ be a square drawn on the segment PE. Where is the point A^2? This depends on the choice of I. If I coincides with A, then $A^2 = -E$. If I coincides with B, then

$$A^2 = (B+E)^2 = B^2 + 2BE + E^2 = -E + 2B + E = 2B.$$

Of course, other choices are also possible, giving other answers.

Among all these possibilities we now choose the one where $I = A$, i.e., I is obtained from E by a rotation through 90° in the positive direction (counterclockwise), and we study it in more detail in the next section.

7. Complex numbers

The points of the line PE are identified with real numbers. Now that we have introduced algebraic operations for the points of the plane, we can view the set of all points as a number system which is wider than real numbers. These numbers are called *complex numbers*. In the conventional notation for complex numbers, our pole P is denoted by 0, point E by 1, point I by i or $\sqrt{-1}$, and $a+bi$ is written instead of $aE+bI$. Here are, once again, the definitions for algebraic operations on complex numbers in this standard notation:

$$\begin{aligned}
(a+bi)+(c+di) &= (a+c)+(b+d)i,\\
(a+bi)-(c+di) &= (a-c)+(b-d)i,\\
(a+bi)(c+di) &= (ac-bd)+(ad+bc)i,\\
\frac{a+bi}{c+di} &= \frac{ac+bd}{c^2+d^2}+\frac{bc-ad}{c^2+d^2}i.
\end{aligned}$$

To put it shortly, the operations are performed as if on polynomials in the "variable" i with the rule $i^2 = -1$ applied whenever possible. To derive the formula for the quotient from this rule, both numerator and denominator should be multiplied by the same number, $c - di$.

The two basic complex numbers 1 and i are referred to as the *real unit* and the *imaginary unit*, respectively.

Exercise 18. Perform the operations on complex numbers:

(a) $\dfrac{3+5i}{1-i} - i(3+i) + \dfrac{1}{i}$,

(b) $\sqrt{3-4i}$,

(c) $\left(\frac{1}{2}-\frac{\sqrt{3}}{2}i\right)^{2004}$.

Let $z = a + bi$. The distance between the points z and 0 is called the *modulus*, or *absolute value*, of the complex number z, and is denoted by $|z|$. Since a and b are Cartesian coordinates of the point z, we have $|z| = \sqrt{a^2+b^2}$. For example, the modulus of both $\cos t + i \sin t$ and $\frac{2t}{1+t^2} + \frac{1-t^2}{1+t^2}i$ is 1 for any value of the real number t.

The distance between the two points represented by complex numbers z and w is $|z - w|$, because the four points 0, w, z, $z - w$ form a parallelogram (see Figure 18).

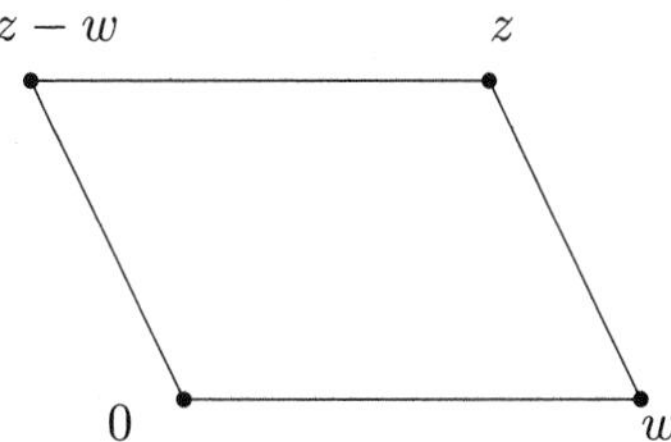

Figure 18. A complex parallelogram

Exercise 19. Find the set of all points z in the complex plane which satisfy:

(a) $|z + 3| = 5$,
(b) $|z + 4| = |z - 2i|$,
(c) the sum of the squares of the distances from z to two fixed points is a given number.

The fact that $\sqrt{a^2+b^2}$ is the distance between two points provides a means to visualize certain purely algebraic problems.

Problem 8. *Prove the inequality*

$$\sqrt{a_1^2+b_1^2}+\sqrt{a_2^2+b_2^2}+\cdots+\sqrt{a_n^2+b_n^2}$$
$$\geq \sqrt{(a_1+a_2+\cdots+a_n)^2+(b_1+b_2+\cdots+b_n)^2}.$$

Solution. Put $z_1 = a_1 + b_1 i$, $\ldots$, $z_n = a_n + b_n i$ and consider the broken line with vertices at 0, z_1, $z_1 + z_2$, $\ldots$, $z_1 + z_2 + \cdots + z_n$. The left-hand side of the inequality is the total length of this line, while the right-hand side is the distance between its endpoints.

Exercise 20. Prove the inequality

$$\sqrt{x_1^2 + (1 - x_2)^2} + \sqrt{x_2^2 + (1 - x_3)^2} + \cdots + \sqrt{x_{10}^2 + (1 - x_1)^2} > 7$$

for any real numbers x_1, ..., x_{10}.

The angle by which the half line 01 should be rotated counterclockwise in order for it to pass through the point z is called the *argument* of the complex number z; it is denoted by $\arg z$. Here 0 and 1 are the points that correspond to the numbers 0 and 1.

Exercise 21. Find the arguments of the following complex numbers: 2, i, -3, $-2i$, $1 + i$, $\sqrt{3} - i$.

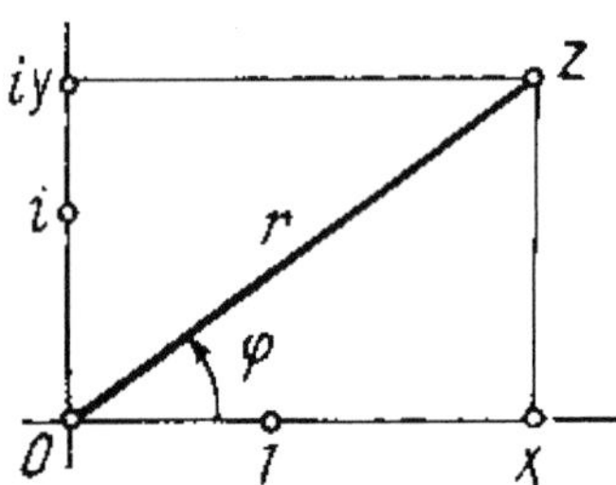

Figure 19. Polar coordinates

A complex number is completely defined if one knows its modulus r and argument φ. Indeed, as you can see in Figure 19, $z = x + yi$, where $x = r\cos\varphi$, $y = r\sin\varphi$; thus

$$z = r(\cos\varphi + i\sin\varphi).$$

This expression is called the *trigonometric form* of the complex number.

The correspondence $z \leftrightarrow (r, \varphi)$ between complex numbers and pairs of real numbers is not one-to-one. For one thing, the argument of the number 0 is undefined. On the other hand, the argument of any non-zero complex number is only defined up to a whole number of complete rotations. Thus, one is free to choose 0, 2π, -2π, 4π, $\ldots$ as the argument of the number 1. Nevertheless, the pair (r, φ) is usually viewed as a pair of coordinates for the point z, called *polar coordinates*.

These coordinates are widely used in practice, e.g., in airport control centres: to determine the location of an aircraft, you first find the direction and then measure the distance.

The equations of some figures look much simpler when written in polar coordinates.

Exercise 22. (a) Plot the line given in polar coordinates by the equation $r = |\cos 3\varphi|$. (b) Find a polar equation which describes a flower with six petals similar to the one shown in Figure 20. Try to rewrite it in Cartesian coordinates.

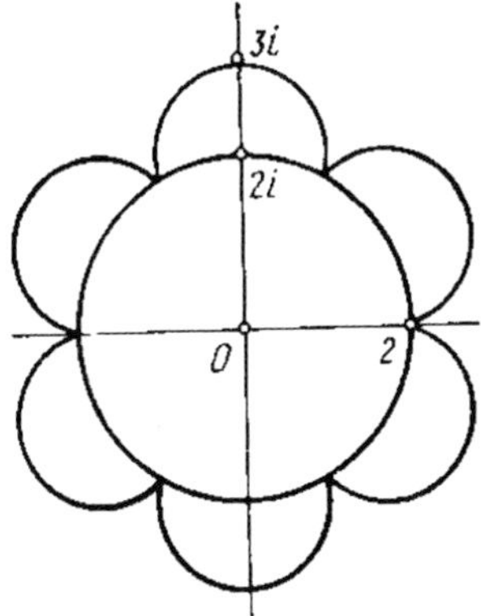

Figure 20. A flower in the complex plane

Multiplication of complex numbers looks simpler when written in terms of modulus and argument. In fact, the following two relations hold:

$$|zw| = |z||w|, \tag{4}$$

$$\arg(zw) = \arg z + \arg w. \tag{5}$$

The first one is a consequence of the remarkable identity

$$(ac - bd)^2 + (ad + bc)^2 = (a^2 + b^2)(c^2 + d^2).$$

To prove the second one, let

$$z = r(\cos\varphi + i\sin\varphi), \qquad w = s(\cos\psi + i\sin\psi).$$

Then

$$\begin{aligned} zw &= rs(\cos\varphi + i\sin\varphi)(\cos\psi + i\sin\psi) \\ &= rs((\cos\varphi\cos\psi - \sin\varphi\sin\psi) + i(\sin\varphi\cos\psi + \cos\varphi\sin\psi)), \end{aligned}$$

which simplifies to

$$zw = rs(\cos(\varphi + \psi) + i\sin(\varphi + \psi)).$$

This proof is based on the well-known trigonometric formulas for the sine and cosine of the sum of two numbers. We will give another, more elegant, proof which only relies on elementary Euclidean geometry and, by the way, implies the trigonometric rules used above.

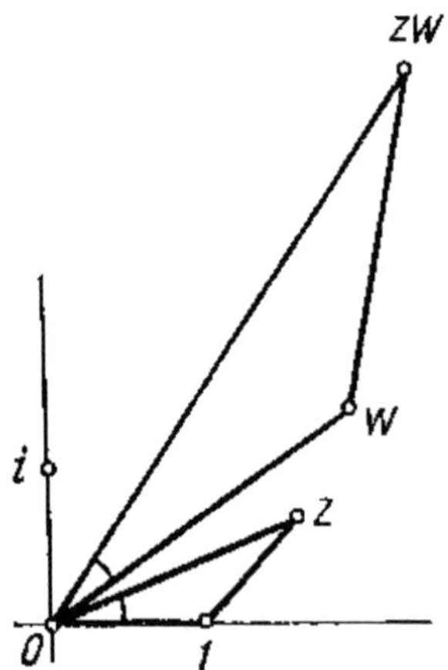

Figure 21. Product of complex numbers

Consider two triangles with vertices 0, 1, z and 0, w, zw (Figure 21). Since $|w| : 1 = |zw| : |z| = |zw - w| : |z - 1|$, these two triangles are similar, so that their respective angles are equal. The equality of the two angles which are marked in Figure 21 proves that $\arg(zw) = \arg z + \arg w$.

OK. Multiplication of complex numbers means that their moduli get multiplied, while their arguments are added. Iterated as appropriate, this observation yields the formula for powers of a complex number in trigonometric notation:

$$[r(\cos\varphi + i\sin\varphi)]^n = r^n(\cos n\varphi + i\sin n\varphi).$$

Exercise 23. Prove that, if z is a complex number and α is a real number such that $z + 1/z = 2\cos\alpha$, then $z^n + 1/z^n = 2\cos n\alpha$.

The trigonometric power formula is very convenient in problems such as Exercise 18c, which you might have already tried. Let us do it together now. Denote $\frac{1}{2} - \frac{\sqrt{3}}{2}i$ by ζ. Then $|\zeta| = 1$ and $\arg\zeta = -\pi/6$; hence $|\zeta^{1998}| = 1^{1998} = 1$ and $\arg\zeta^{1998} = 1998 \cdot (-\pi/6) = -333 \cdot 2\pi$. This implies that $\zeta^{1998} = 1$.

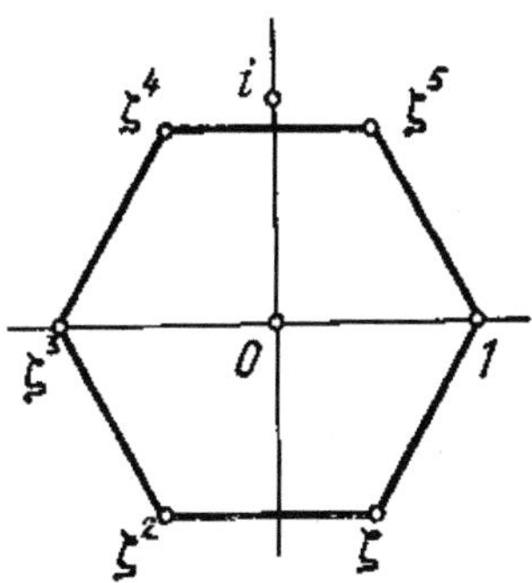

Figure 22. Complex roots of 1

Note that an integer power of the number ζ can occupy only one of the six positions in the plane — the vertices of the regular hexagon shown in Figure 22. Any of these six complex numbers is a power of ζ and plays the role of a sixth root of the number 1, because $(\zeta^k)^6 = (\zeta^6)^k = 1^k = 1$. In general, for any natural n, there are exactly n complex n-th roots of unity, arranged as the vertices of a n-gon.

Problem 9. *Find the product of all diagonals and both sides that issue from one vertex of the regular n-gon inscribed into the circle of radius 1.*

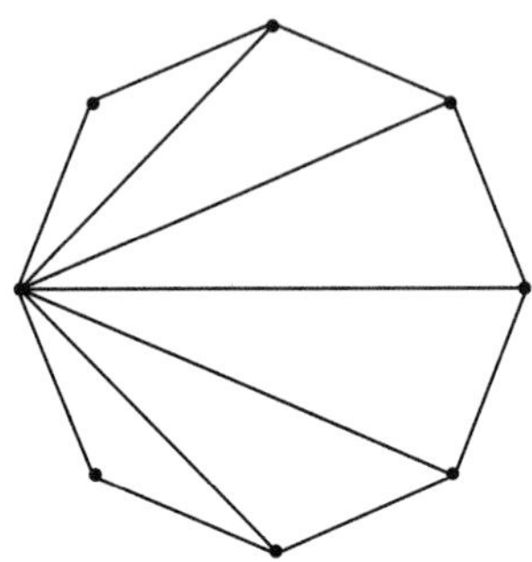

Figure 23. Sides and diagonals of a regular polygon

Solution. Put the pole (number 0) at the centre of the polygon and the real unity (number 1) at the given vertex A_1. All the vertices are the roots of the equation $z^n - 1 = 0$; therefore all the vertices but A_1 also satisfy the equation $z^{n-1}+z^{n-2}+\cdots+z+1=0$ obtained by dividing z^n-1 by $z-1$. Now compare the two polynomials $z^{n-1}+z^{n-2}+\cdots+z+1=0$ and $(z-A_2)(z-A_3)\ldots(z-A_n)$. They are identically equal, because they have the same roots and equal leading coefficients. Hence, their values at $z=A_1$ are equal:

$$(A_1-A_2)(A_1-A_3)\ldots(A_1-A_n)=A_1^{n-1}+A_1^{n-2}+\cdots+A_1+1.$$

Recalling that $A_1=1$ by our choice, we obtain the answer:

$$|A_1-A_2||A_1-A_3|\ldots|A_1-A_n|=n.$$

Exercise 24. A regular polygon $A_1A_2\ldots A_n$ is inscribed in the circle of unit radius, and A is an arbitrary point of this circle. Find the sum of squares of distances from A to all the vertices of the polygon.

Since division is an operation inverse to multiplication, it satisfies the formulas inverse to (4):

$$\left|\frac{z}{w}\right| = \frac{|z|}{|w|},$$

$$\arg\frac{z}{w} = \arg z - \arg w.$$

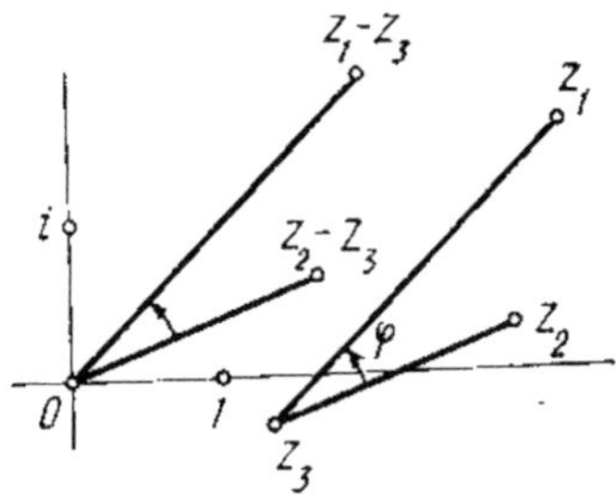

Figure 24. Angle expressed through complex numbers

The latter equality is interesting from the point of view of elementary geometry: it allows us to express the magnitude of an angle in terms of its vertex and two points belonging to its sides:

$$\varphi = \arg\frac{z_1 - z_3}{z_2 - z_3}$$

(Figure 24). Here are two examples where this observation is applied: in the first one, we solve a geometric problem using the algebra of complex numbers, and in the second one, conversely, we solve an algebraic problem by a geometric method.

Problem 10. *Three squares are placed side by side as shown in Figure 25. Prove that the sum of $\angle KAH$, $\angle KDH$ and $\angle KFH$ is a right angle.*

Solution. Evidently, $\angle KFH = \pi/4$, so we have to prove that $\angle KAH + \angle KDH = \pi/4$, too. Assuming that $A = 0$, $D = 1$ and $B = i$, we have $F = 2$, $K = 3$, $H = 3 + i$. Therefore, $\angle DAH = \arg\frac{H-A}{D-A} = \arg(3 + i)$,

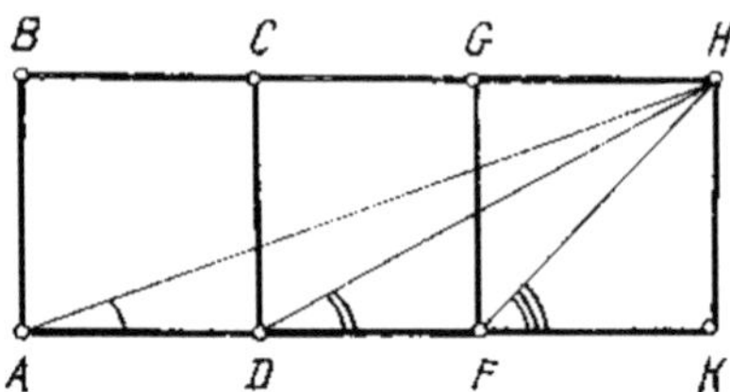

Figure 25. Sum of three angles

$\angle FDH = \arg \frac{H-D}{F-D} = \arg(2+i)$, whence $\angle DAH + \angle FDH = \arg(3+i)(2+i) = \arg(5+5i) = \pi/4$, which is just what was required.

Problem 11. *Prove that if z_1, z_2, z_3, z_4 are different complex numbers with equal absolute values, then*

$$\frac{z_1 - z_3}{z_2 - z_3} : \frac{z_1 - z_4}{z_2 - z_4}$$

is a real number.

Solution. The four given points lie on the same circle centred at 0. Points z_1 and z_2 split this circle into two arcs. The other two points z_3, z_4 can belong either to the same arc, or to different arcs. In the first case the angles $z_1z_3z_2$ and $z_1z_4z_2$ are equal, because they subtend the same arc. Therefore, $\arg \dfrac{z_1 - z_3}{z_2 - z_3} = \arg \dfrac{z_1 - z_4}{z_2 - z_4}$ and $\arg \dfrac{z_1 - z_3}{z_2 - z_3} : \dfrac{z_1 - z_4}{z_2 - z_4} = 0$, i.e. the number in question is real and positive. In the second case the two angles $z_1z_3z_4$ and $z_2z_4z_1$ have the same orientation and together make $180°$. Therefore, the number in question is real and negative.

The assertion of Problem 11 evidently generalizes to any set of four complex numbers that belong to an arbitrary circle or straight line in the plane. The converse is also true: if the given expression is real, then the four numbers must belong either to the same circle or to the same straight line.

Exercise 25. Let $c_1, c_2, \dots, c_n$ be the vertices of a convex polygon. Prove that all complex roots of the equation

$$\frac{1}{z-c_1} + \frac{1}{z-c_2} + \cdots + \frac{1}{z-c_n} = 0$$

are interior points of this polygon.

Chapter 2

Plane Movements

Plane movements are transformations of the plane that do not change the lengths of segments and, as a consequence, preserve all parameters of geometric figures, such as areas, angles, etc.

We begin this chapter with the discussion of some well-known problems of elementary geometry that allow a short solution using plane movements. All these problems share the same underlying idea: change the position of certain parts of the given geometric configuration in such a way that the hidden relations between the elements become transparent.

We then proceed to a detailed discussion of the composition of movements, which will provide experimental material for the introduction of transformation groups in the next chapter.

1. Parallel translations

Definition 4. A *parallel translation* (or simply a *translation*) is a transformation of the plane that sends every point A into the point A' such that $\overrightarrow{AA'}$ is equal to a given constant vector $\mathbf{v}$. This transformation is denoted by $T_{\mathbf{v}}$.

Problem 12. *Two villages A and B are located across the river from each other. The sides of the river are rectilinear and parallel to each other. Where should one build the bridge MN so that the distance*

$AMNB$ be as small as possible? The bridge must be perpendicular to the sides of the river.

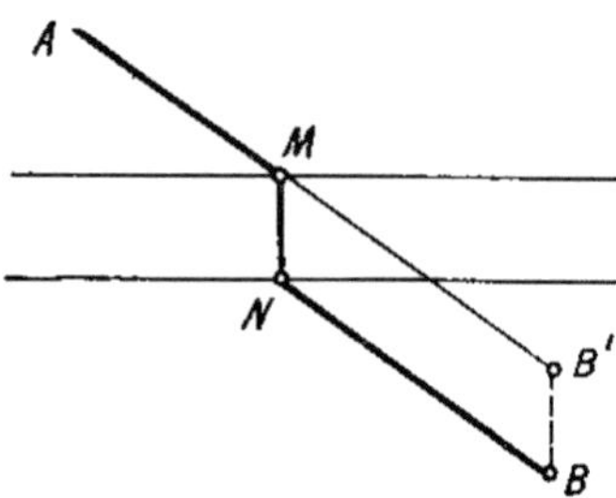

Figure 1. Bridge over a river

Solution. If there were no river, the shortest path joining A and B would be a straight line. Let us try to get rid of the river by moving one of its sides towards the other perpendicularly until both sides coincide (Figure 1). Let B' be the new position of the point B. The lengths of $AMB'B$ and $AMNB$ are equal. The position of point B' does not depend on the choice of the place for the bridge. Hence we only have to minimize the distance AMB', which is can be done simply by making AMB' a straight line.

Exercise 26. Construct the shortest path that connects two points A and B separated by two rivers (Figure 2). Both bridges must be perpendicular to the sides of the rivers.

Problem 13. *Inscribe a given vector in a given circle (i.e., construct a chord of a given circle which is equal and parallel to a given segment).*

Solution. Let AB be the given vector and C the given circle with centre O and radius r (Figure 3). We have to move AB, keeping it parallel to itself, towards C so that it gets inscribed into the circle. In fact, it is much

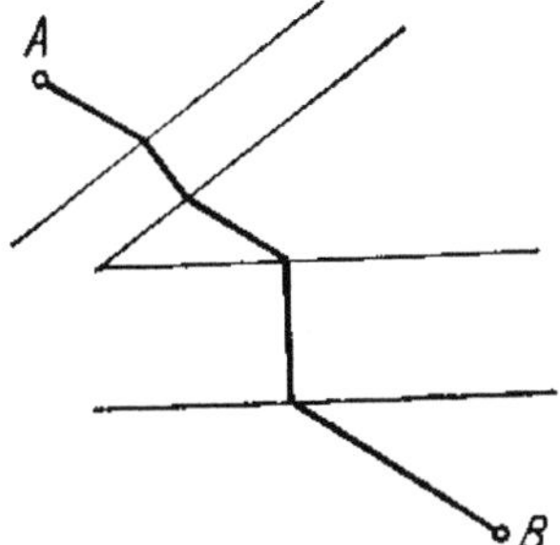

Figure 2. Two bridges over two rivers

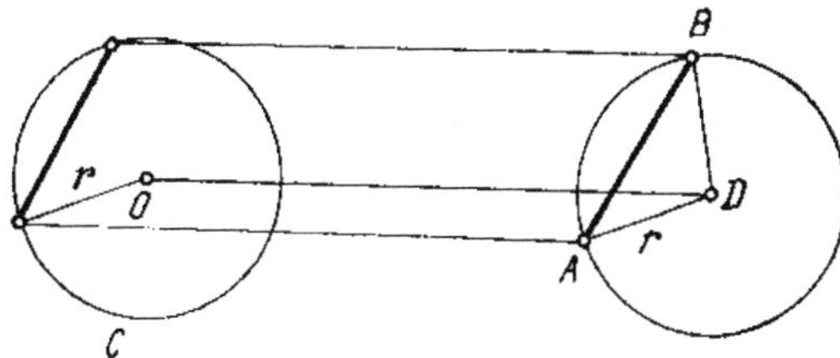

Figure 3. Inscribing a vector into a circle

easier to perform the reverse operation: move the circle in the opposite direction so that in the new position it will pass through both endpoints of the vector, A and B. To do so, we construct the triangle ABD such that $AD = BD = r$. The point D is the centre of the moved circle. Now if we translate the points A and B by the vector $\overrightarrow{DO}$, we will obtained the segment inscribed into the initial circle.

Here are two more problems which can be solved using parallel translation.

Exercise 27. Inscribe a given vector in a given triangle, i.e., find a segment whose endpoints lie on the sides of the given triangle and which is equal and parallel to a given segment.

Exercise 28. Construct a trapezoid if the lengths of its parallel sides and diagonals are known.

2. Reflections

Definition 5. Let l be a line in the plane. The *reflection with respect to* l is a transformation of the plane that sends every point A into the point A' such that l is the perpendicular bisector of the segment AA'. This transformation is denoted by S_l. and is also called *axial symmetry with axis* l.

Problem 14. *Two points* A *and* B *are on one side of the straight line* l. *Find the point* $M \in l$ *such that the length of the broken line* AMB *is minimal. If you prefer 'real life' problems, you may imagine a person with an empty bucket at point* A, *a fire at point* B *and a straightline river* l.

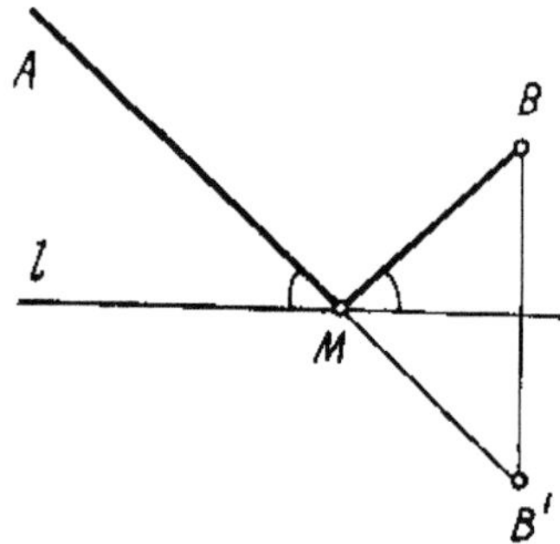

Figure 4. Shortest path

Solution. If both points A and B were situated on different sides of the line l, the solution would be a straight line AB. Let us try to reduce our problem to this case by reflecting the given point B in the line l (see Figure 4). If B' is the image of B, then the lines AMB and AMB' have equal lengths for any arbitrary position of the point $M \in l$. To minimize this distance, we draw the straight line AB' and set M to be the intersection point of this line with l. Note that in this case the angles formed by either of the two lines AM and BM with l are the same, which agrees with the well-known law of optics.

Exercise 29. Inside an angle XOY, two points, A and B, are given. Among all broken lines $AMNB$ where $M \in XO$, $N \in YO$, find the line of minimal length[1].

Figure 5. Two rivers of Exercise 29

We proceed with one more problem related to shortest paths.

Problem 15. *Into a given acute triangle inscribe a triangle of minimal perimeter.*

Solution. Let UVW be an arbitrary triangle inscribed in the given triangle ABC. Let K and L be the symmetric images of the point U with respect to the lines AB and

[1]Figure 5 refers to a Russian folk tale where a raven has to bring two kinds of water, the 'dead' water and the 'live' water, to revive the prince.

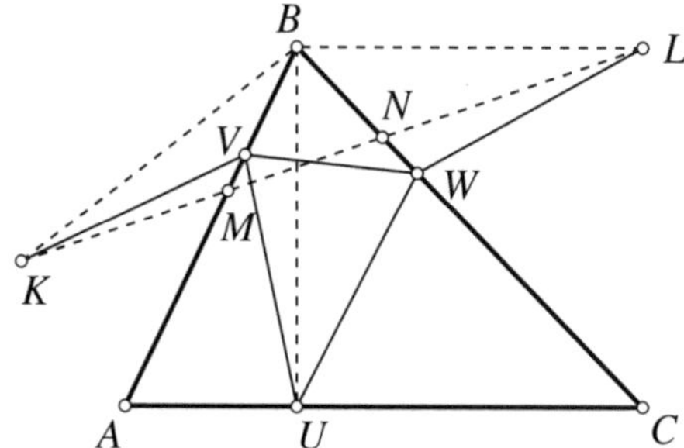

Figure 6. Inscribed triangle of minimal perimeter

BC (see Figure 6). The paths $UVWU$ and $KVWL$ have equal lengths. To minimize this length among all triangles UVW with a fixed vertex U, we have to choose V and W so that $KVWL$ becomes a straight line, i.e. to set $V = M$ and $W = N$. Now among all triangles $\triangle UMN$ that correspond to different positions of the point U, we will choose the one with the minimal perimeter, and it will give the solution to the problem. We have to find the position of U for which the segment KL is shortest.

Note that $\triangle BKL$ is an isosceles triangle with $BK = BU = BL$. Its angle at vertex B does not depend on the position of the point U: $\angle KBL = 2\angle ABC$. Therefore, to minimize the length of the side KL we have to make sure that the side BK is as small as possible. Since $BK = BU$, this minimum is attained when U is the base point of the altitude drawn in the triangle ABC from the vertex B: $BU \perp AC$.

Because of the symmetry between the three points U, V and W, we conclude that V and W in the minimal triangle UVW are also basepoints of the corresponding altitudes of the triangle ABC.

Exercise 30. Construct a triangle, if one of its vertices and the three lines that contain its bisectors are given.

Exercise 31. A ray of light enters an angle of 45° formed by two mirrors. Prove that after several reflections the ray will exit the angle moving along a line parallel to its initial trajectory. Are there other values of the angle with the same property?

3. Rotations

Definition 6. Let O be a point in the plane and φ a real number, understood as an angle. The *rotation around O through angle φ* is a transformation of the plane that sends every point A into the point A' such that $|OA| = |OA'|$ and $\angle AOA' = \varphi$, where the angle is counted with sign, the counterclockwise direction being considered as positive. This transformation is denoted by R_O^φ.

Look at Figure 7. It is evident that the sum of all vertices of a regular polygon with an even number of vertices over its centre P is equal to P (see p. 10 for the definition of point addition). Indeed, the set of vertices splits into pairs of mutually opposite points. It is not so easy to prove the same property for a polygon with an odd number of vertices. If you try to directly compute the coordinates of all the vectors, you will have to deal with rather complicated trigonometric expressions. However, the problem looks difficult only as long as the plane does not move.

Problem 16. *Prove that the sum of vertices of a regular polygon over its centre P coincides with P.*

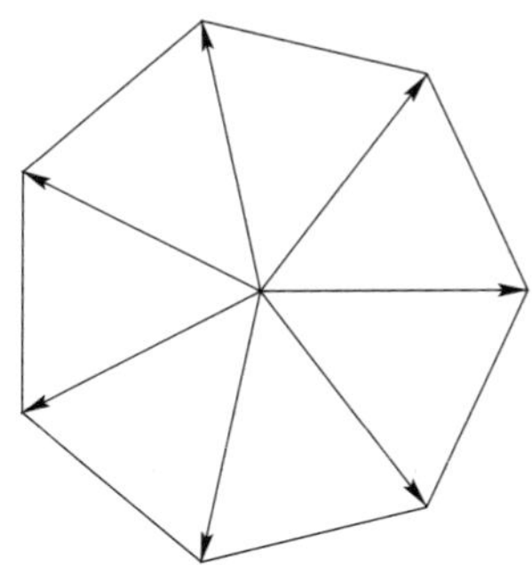

Figure 7. Sum of vertices of a polygon

Solution. Let n be the number of vertices. Under a rotation through $360/n$ degrees around P the given polygon goes into itself. Therefore, the sum of vertices remains unchanged. But in the plane there is only one point that goes into itself under a rotation: it is the centre of the rotation.

Exercise 32. A point M lies inside a convex polygon. Perpendiculars are drawn from M to all sides of the polygon, and on each of these half-lines, a point A_i is taken whose distance from M equals the length of the corresponding side. Prove that the sum of all these points over M is zero.

Problem 17. *Construct an equilateral triangle, if the distances of its vertices from a given point D are a, b and c.*

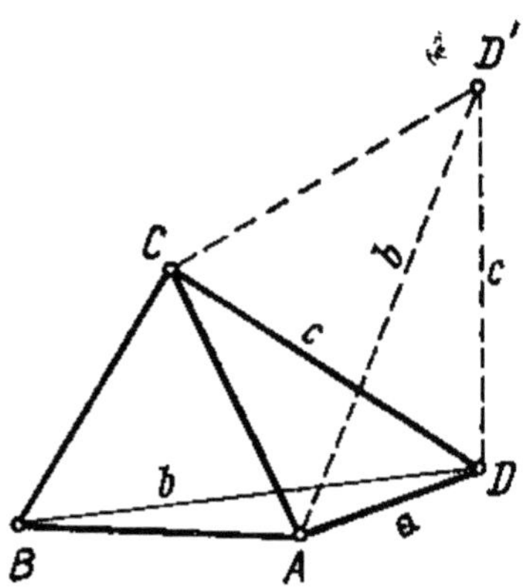

Figure 8. Constructing an equilateral triangle

Solution. Everyone knows how to construct a triangle when given the lengths of its sides. Unfortunately, the three segments a, b, c in Figure 8 do not form a triangle. Let us rotate the plane by $60°$ around the point C. The point B goes into A and D goes into D'. A rotation preserves distances; therefore the lengths of the sides of the triangle $\triangle ADD'$ are a, b and c. We will construct this triangle first, then find the point C ($\triangle CDD'$ is equilateral), and finally find the point B.

Exercise 33. Construct an equilateral triangle whose vertices lie on three given parallel lines, one on each.

Problem 18. *Inside a given triangle, find the point the sum of whose distances from the vertices is minimal.*

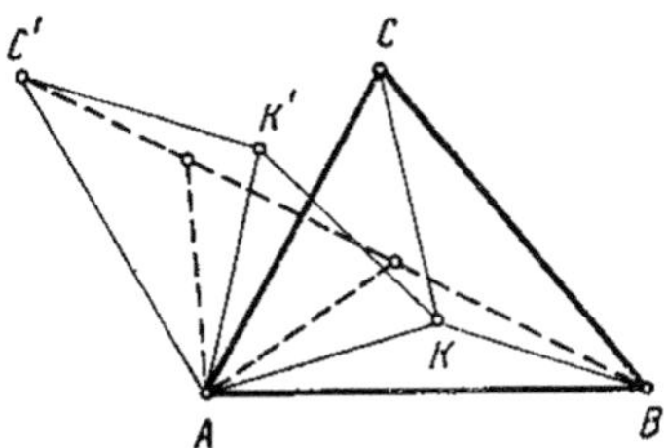

Figure 9. Minimize the sum of distances

Solution. Let K be an arbitrary point inside the triangle ABC. Rotate the points C and K around A counterclockwise through $60°$ and denote their new positions by C' and K' (see Figure 9). The sum of the three distances in question, $AK + BK + CK$, equals the length of the broken line $C'K'KB$. It is minimal if K and K' lie on the straight line BC'. Thus, the optimal position for K is the point K_0 on BC' such that the angle AK_0C' is $60°$ or, in other words, $\angle AK_0B = 120°$. By symmetry, we also have $\angle BK_0C = \angle CK_0A = 120°$.

Note that our analysis, as well as the answer, holds only for triangles whose angles are smaller than $120°$. We leave it to the reader to guess the answer in the opposite case.

Exercise 34. M is an arbitrary point inside a square $ABCD$. Draw four lines which pass through A, B, C and D and are perpendicular to BM, CM, DM and AM, respectively. Prove that these four lines pass through a common point.

The rotation through 180° is also referred to as *half turn*, or *central symmetry*. Speaking about central symmetries, we will often leave 180° out of the notation, writing R_A instead of $R_A^{180^\circ}$. Here are two problems where this kind of movements is used.

Exercise 35. Through the intersection point of two circles, draw a line on which these circles cut equal chords.

Exercise 36. There is a round table and an unlimited number of equal round coins. Two players take turns at placing the coins on the table in such a way that they do not touch each other. What is the winning strategy for the first player?

4. Functions of a complex variable

We return once again to Problem 16 (see page 47). Apart from the geometric solution given above, this problem also has an algebraic solution. To explain it, we introduce a *complex structure* in the plane. More precisely, we choose a one-to-one correspondence between complex numbers and points in the plane in such a way that 0 corresponds to the centre of the polygon and 1 corresponds to one of its vertices. If ζ is the vertex adjacent to 1 (in the counterclockwise direction), then the remaining vertices are $\zeta^2, \ldots, \zeta^{n-1}$. We are interested in $x = 1 + \zeta + \zeta^2 + \cdots + \zeta^{n-1}$. Since $\zeta^n = 1$, we have $x\zeta = \zeta + \zeta^2 + \cdots + \zeta^{n-1} + 1 = x$, which implies that $x = 0$, because $\zeta \neq 1$.

Note that this algebraic proof is essentially the same as the geometric proof given above. More exactly, it is nothing but the translation of the geometric argument into algebraic language. Indeed, the new proof is based on the fact that the only number which satisfies the equation $\zeta x = x$ is $x = 0$. But what happens with a complex number when it is multiplied by ζ? According to the general rule, its modulus remains the same, because $|\zeta| = 1$, and its argument

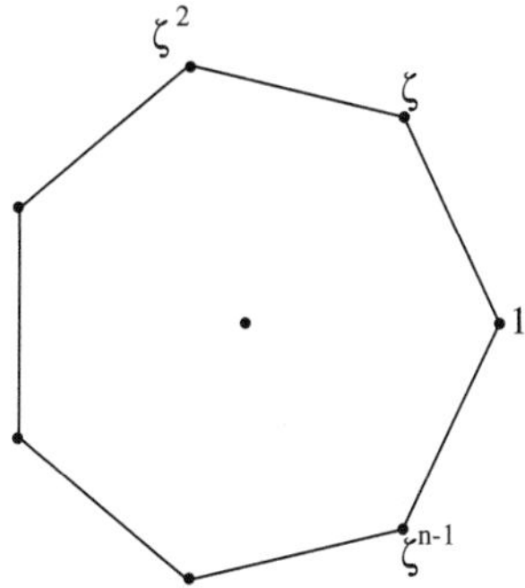

Figure 10. Regular polygon with complex vertices

increases by $360°/n$. In geometric terms, this means that the corresponding point rotates through $360/n$ degrees around the centre of the polygon.

In general, if points are viewed as complex numbers, then transformations of the plane, and in particular, plane movements, should be understood as *functions of a complex variable* $w = f(z)$, where z denotes an arbitrary point and w its image. For example, a rotation around 0 is represented by the function $w = \alpha z$, where $|\alpha| = 1$ (we have in this case $\alpha = \cos\varphi + i\sin\varphi$, where φ is the angle of rotation). It is likewise evident that the formula for a parallel translation is

$$w = z + a, \tag{6}$$

where a is a certain complex number.

Now let us derive the formula for the rotation of the complex plane around an arbitrary point p. Figure 11 shows that the rotation of the point z around p through an angle φ can be split into three steps:

(1) translation $z \mapsto z - p$;

(2) rotation around the origin $z - p \mapsto \alpha(z - p)$;

(3) inverse translation $\alpha(z - p) \mapsto \alpha(z - p) + p$.

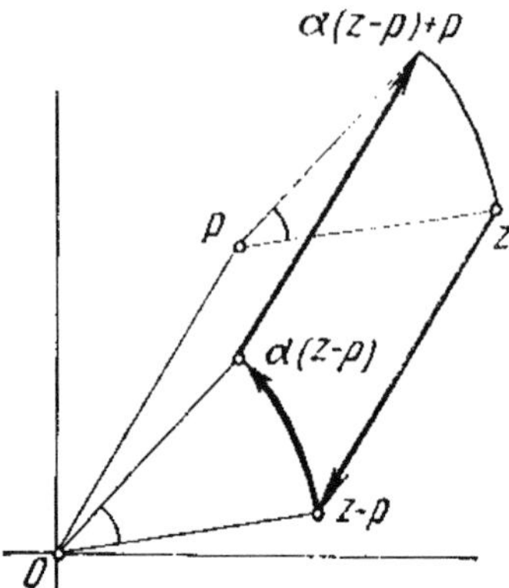

Figure 11. Plane movement in complex coordinates

The rotation around p through the angle φ is thus described by the function

$$w = \alpha z + (1 - \alpha)p, \tag{7}$$

where $\alpha = \cos\varphi + i\sin\varphi$.

Parallel translations and rotations are referred to as *proper movements.* This expression is accounted for by the fact that one does not have to leave the plane in order to physically effectuate one of these transformations, whereas a reflection in a line requires a rotation (flipping) of the plane in the surrounding three-space.

Theorem 3. *The set of proper movements of the plane coincides with the set of all transformations described by the functions of a complex variable*

$$w = \alpha z + m, \tag{8}$$

where α and m are complex numbers and $|\alpha| = 1$.

Proof. Formulas (6) and (7) imply that any proper movement of the plane is described by a linear functions of type (8).

We will prove that the converse also holds, i.e. that every function (8) defines a proper movement. Indeed, if $\alpha = 1$, then (8) becomes

(6) and we deal with a parallel translation. If $\alpha \neq 1$, then (8) can be rewritten as

$$w = \alpha z + m = \alpha\left(z - \frac{m}{1-\alpha}\right) + \frac{m}{1-\alpha},$$

which is the expression of the rotation around $p = m/(1-\alpha)$ through the angle φ such that $\cos\varphi + i\sin\varphi = \alpha$. □

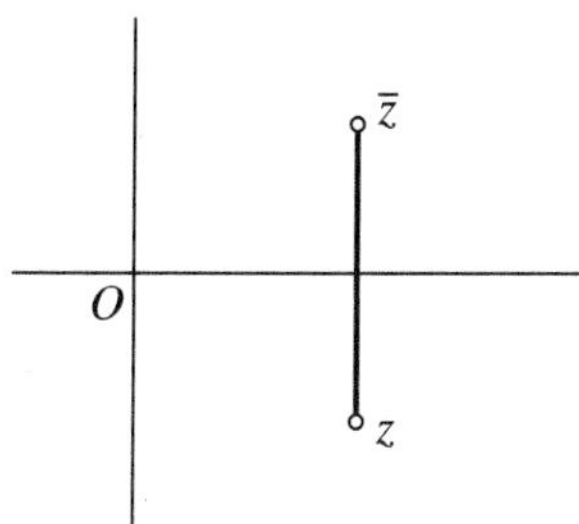

Figure 12. Complex conjugation

To find a similar description for the improper movements, for example reflections, apart from addition and multiplication of complex numbers, we need one more operation: *complex conjugation*. The conjugate of the number $z = x + iy$ is defined as $\bar{z} = x - iy$. Geometrically, conjugation corresponds to reflection in the real axis (Figure 12). Recall that we have already used conjugation to derive the formula for the quotient of two complex numbers (see p. 30).

Exercise 37. Prove the following formulas for reflection in the line $y = kx + b$:

(9) $$w = \bar{z} + 2bi, \quad \text{if} \quad k = 0,$$

(10) $$w = \frac{1 - k^2 + 2ki}{1 + k^2}\left(\bar{z} + \frac{b}{k}\right) - \frac{b}{k}, \quad \text{if} \quad k \neq 0$$

(note that $(1 - k^2 + 2ki)/(1 + k^2) = \alpha^2$, where $\alpha = \cos\varphi + i\sin\varphi$ and φ is the angle between the given line and the x-axis).

In the following example we use the algebra of complex numbers to solve a geometric problem.

Problem 19. *A pirate is hunting for a hidden treasure. According to a letter he has got, he has to go to the Treasure Island, find two* *trees A and B, a rock C (Figure 14) and dig for the buried treasure at the point K which is the middle point of the segment DE, where D is obtained by rotating C around A clockwise through $90°$, and E is obtained by rotating C around B counterclockwise through $90°$. When the pirate arrived at this place, he found that the trees A and B are there, but the rock C disappeared. Is it still possible to recover the position of point K?*

Figure 13. A pirate

Solution. Let us introduce a complex structure in the plane, i.e., associate the points with complex numbers, in such a way that A corresponds to 0, while B and C correspond to numbers b and c (see Figure 14).

Then, by formula (7), points D and E are represented by the numbers $-ic$ and $i(c-b)+b$; therefore, point K is $\frac{1-i}{2}b$. As this expression does not involve c, we see that the position of the hiding place does not depend on the choice of the point C. We also see that K is the vertex of the isosceles right triangle with hypotenuse AB, and as such can be found by our treasure hunter.

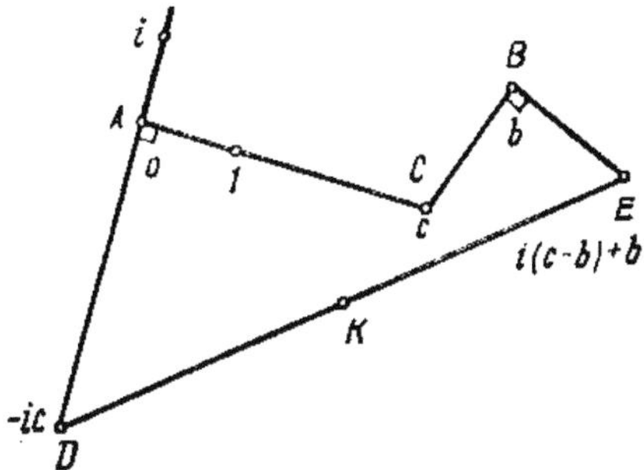

Figure 14. Where is the treasure?

Exercise 38. Two sides of a triangle are rotated through $90°$ around their common vertex in opposite directions. Prove that the line joining the new endpoints is perpendicular to the median of the triangle.

5. Composition of movements

Given two movements of the plane, f and g, one can construct a third movement $g \circ f$, the *composition*, or the *product* of the given two, by performing first f, then g.

Definition 7. The *composition* $f \circ g$ of two movements f and g is defined by the relation

$$(f \circ g)(x) = f(g(x))$$

for any point x.

The transformation $f \circ g$ thus defined is really a movement, because it evidently preserves the distances between the points. In this section, we will study the composition of special types of movements: translations, reflections and rotations.

Problem 20. *Find the composition of two reflections.*

Solution. Let S_l denote the reflection in the line l. Suppose that two lines, l and m, are given, and we have to find the composition $S_m \circ S_l$. Let A' be the image of an arbitrary point A under the movement S_l, and A'' the image of A' under S_m.

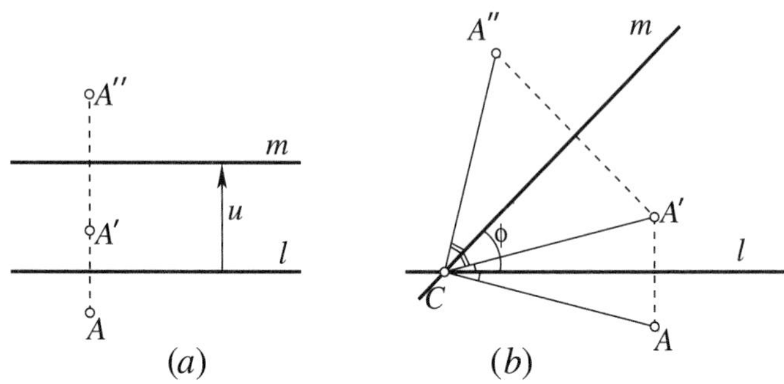

Figure 15. Product of two reflections

We first consider the case when the two lines l and m are parallel to each other (Figure 14a). Then all the three points A, A', A'' lie on one line, perpendicular to l and m, and the distance between the points A and A'' is twice the distance between the lines l and m, independent of the position of A. Therefore, the composition of the two reflections S_l and S_m has the same effect as translation by the vector $2\mathbf{u}$, where $\mathbf{u}$ is the vector perpendicular to l and m, of length equal to the distance between the two lines and directed from l to m:

(11) $$S_m \circ S_l = T_{2\mathbf{u}}.$$

Now suppose that the lines l and m meet at a certain point C (Figure 14b). If φ is the angle between l and m, then, as you can see from the figure, $\angle ACA'' = 2\varphi$. Note also that all the three points A, A' and A'' are at the same distance from C. It follows that

(12) $$S_m \circ S_l = R_C^{2\varphi},$$

where $R_C^{2\varphi}$ denotes rotation around C through the angle 2φ (clockwise if $\varphi < 0$ and counterclockwise if $\varphi > 0$).

The reader may wish to consider other locations for the point A in the plane, different from that of Figure 15b, and make sure that formulas (11) and (12) are always true. Bear in mind that the angle φ should be measured from line l to line m, i.e., for example, $\varphi = \pi/4$ means that m can be obtained from l by a positive (counterclockwise) rotation through $45°$.

Formula (12) implies, by the way, that the composition of two movements in general depends on the order in which they are taken: thus, $S_l \circ S_m$ is the movement inverse to $S_m \circ S_l$.

Exercise 39. Let l, m and n be three lines meeting at one point. Find the movement $(S_n \circ S_m \circ S_l)^2 = S_n \circ S_m \circ S_l \circ S_n \circ S_m \circ S_l$. We suggest that the reader first experiment by applying the given composition to an arbitrary point of the plane, and then prove the result using the formulas we have established.

Formulas (11) and (12), read from right to left, show how to decompose a translation or a rotation into a product of reflections. This decomposition is not unique, and the freedom we have in the choice of the axes of reflection may prove quite useful for the solution of a specific problem.

Problem 21. *Find the composition of two rotations.*

Solution. If the centres of both rotations coincide, then the answer is obvious:

(13) $$R_A^\varphi \circ R_A^\psi = R_A^{\varphi+\psi}.$$

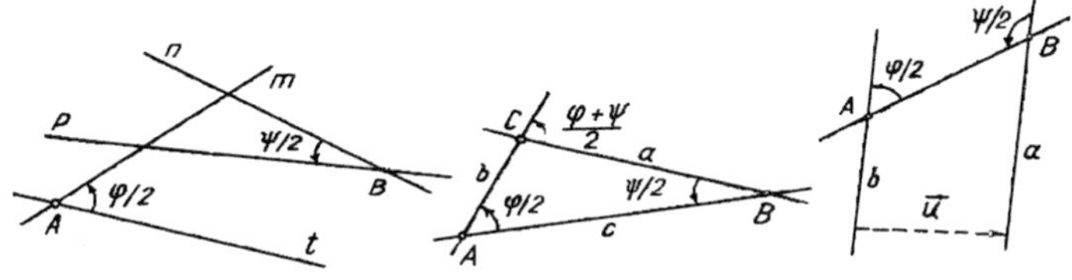

Figure 16. Product of two rotations

Now consider two rotations R_A^φ and R_B^ψ with different centres. To find their composition, we will represent each rotation as the product of two reflections and then use the formulas that we already know. We have $R_A^\varphi = S_m \circ S_l$, where the lines l and m form an angle $\varphi/2$ at the point A, and $R_B^\psi = S_p \circ S_n$, where the lines n and p form an angle $\psi/2$ at the point B (see Figure 16a). Then $R_A^\varphi \circ R_B^\psi = S_m \circ S_l \circ S_p \circ S_n$. This expression simplifies to $S_m \circ S_n$ when the two lines l and p coincide, because in this case $S_l \circ S_p = \mathrm{id}$ is the *identity* transformation, i.e., the transformation which takes every point into itself.

After this analysis, we start anew from Figure 16b. We denote by c the line joining A and B; then, rotating c around A through the angle $\varphi/2$ and around B through $-\psi/2$, we obtain the lines b and a. If the lines b and c have a common point, we denote it by C, and in this case we can write

$$R_A^\varphi \circ R_B^\psi = S_b \circ S_c \circ S_c \circ S_a = S_b \circ S_a = R_C^{\varphi+\psi},$$

or, setting $\alpha = \varphi/2, \beta = \psi/2, \gamma = \pi - \alpha - \beta$,

(14) $$R_A^{2\alpha} \circ R_B^{2\beta} = R_C^{-2\gamma},$$

where C is the third vertex of the triangle with two vertices A and B and angles at these vertices equal to α and β; γ is the angle of this triangle at C.

After both sides of (14) are multiplied by $R_C^{2\gamma}$ on the right, it takes a more symmetric form:

(15) $$R_A^{2\alpha} \circ R_B^{2\beta} \circ R_C^{2\gamma} = \mathrm{id}\,.$$

The converse is also true: if the three points A, B, C and three angles α, β, γ between 0 and 180° satisfy equation (15), then α, β and γ are equal to the angles of the triangle ABC.

Equality (15) can be checked directly. Since $2\alpha + 2\beta + 2\gamma = 360°$, the composition $R_A^{2\alpha} \circ R_B^{2\beta} \circ R_C^{2\gamma}$ is a parallel translation. To prove that it is the identity, we just need to check that it has one fixed point. But Figure 17 shows that the point A remains fixed under the successive mappings $R_C^{2\gamma}$, $R_B^{2\beta}$, $R_A^{2\alpha}$.

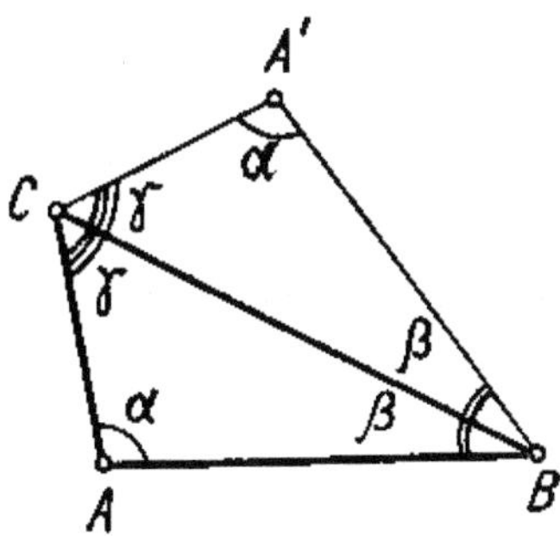

Figure 17. Composition of three rotations

Should the lines a and b be parallel (this happens when $\varphi + \psi$ is a multiple of 2π), then

(16) $$R_A^{2\alpha} \circ R_B^{2\beta} = T_{2\mathbf{u}},$$

where $\mathbf{u}$ is defined in Figure 16c.

Using complex numbers, one can derive an algebraic formula for the composition of two rotations. We take a complex number z and apply successively first the rotation R_B^{ψ}, then the rotation R_A^{φ}. According to formula (7), we can write

$$\begin{aligned} R_B^{\psi}(z) &= q(z-b)+b, \\ R_A^{\varphi}(w) &= p(w-a)+a, \end{aligned}$$

where $p = \cos\varphi + i\sin\varphi$, $q = \cos\psi + i\sin\psi$. Now we substitute $R_B^{\psi}(z)$ instead of w and try to rewrite the result in a similar form:

$$\begin{aligned} (R_A^{\varphi} \circ R_B^{\psi})(z) &= p(q(z-b)+b-a)+a \\ &= pq\big(z - \frac{a-pa+pb-pqb}{1-pq}\big) + \frac{a-pa+pb-pqb}{1-pq}. \end{aligned}$$

Note that $pq = \cos(\varphi+\psi) + i\sin(\varphi+\psi)$. Therefore, the result obtained means that

$$R_A^{\varphi} \circ R_B^{\psi} = R_C^{\varphi+\psi},$$

where the point C corresponds to the complex number

$$c = \frac{a - pa + pb - pqb}{1 - pq}. \tag{17}$$

We see that geometric and algebraic arguments lead to two different formulas for the composition of rotations. We can benefit from this fact by deriving the following corollary:

If two vertices A and B of a triangle ABC correspond to complex numbers a and b, and the angles at these vertices are $\varphi/2$ and $\psi/2$, then the third vertex, as a complex number, is determined by formula (17).

We pass to examples where the composition of movements and the formulas we have found are used.

Problem 22. *Three equilateral triangles are built on the sides of an arbitrary triangle ABC (Figure 18). Prove that their centres M, N, P form an equilateral triangle.*[2]

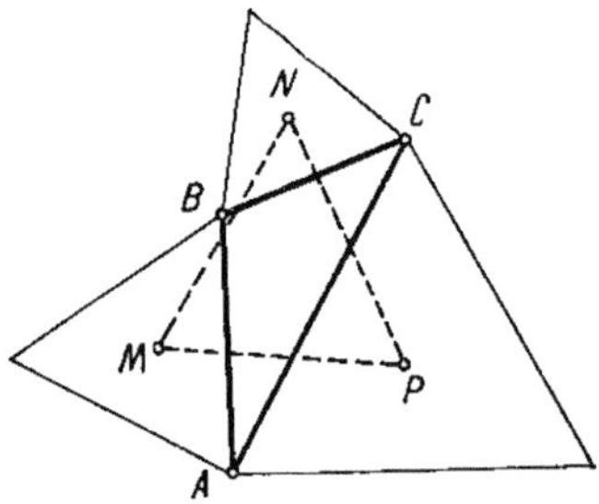

Figure 18. Problem of Napoléon

Solution. Triangles AMB, BNC and CPA are isosceles with obtuse angles of $120°$. Consider the composition of three rotations $F = R_P^{120°} \circ R_N^{120°} \circ R_M^{120°}$. Formulas

[2]This problem is known as *the problem of Napoléon*, although the famous French general is not its author.

(14) and (15) show that F is either a rotation or a parallel translation. Since the sum of the three angles of rotation is 360°, F must be a parallel translation. Let us trace how the point A is moved by F. It is clear that $R_M^{120^\circ}(A) = B$, $R_N^{120^\circ}(B) = C$, $R_P^{120^\circ}(C) = A$, and thus $F(A) = A$. It follows that F is a translation by zero vector, i.e.,

$$R_P^{120^\circ} \circ R_N^{120^\circ} \circ R_M^{120^\circ} = \mathrm{id}\,.$$

Comparing this to (15), we conclude that M is the third vertex of the triangle having two vertices at N and P and angles 60° and 60° at these vertices, i.e. an equilateral triangle.

Exercise 40. Find a solution of the previous problem based on computations with complex numbers.

Exercise 41. On the sides of an arbitrary quadrangle four squares are built. Prove that their centres form a quadrangle whose diagonals are mutually perpendicular and have equal length.

Exercise 42. Find the composition of

1. two central symmetries,
2. a central symmetry and a reflection.

Exercise 43. Construct a pentagon, given the midpoints of all its sides.

6. Glide reflections

We have studied three types of plane movements: translations, rotations and reflections. However, these three types do not cover all plane movements. For example, in Exercise 42, the product of a reflection and a central symmetry does not belong to any of these types.

Definition 8. A *glide reflection with axis l and vector v* is a movement that consists in a reflection with respect to a line l and a translation by the vector v, which is assumed to be parallel to the line l (see Figure 19).

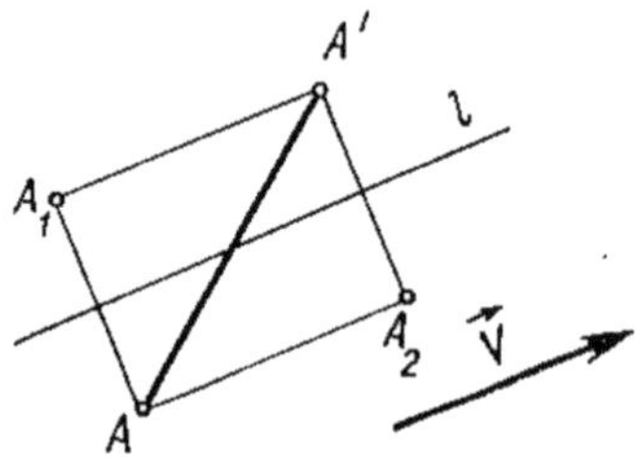

Figure 19. Glide reflection

Denoting the glide reflection by U_l^v, we can write the definition as $U_l^v = T_v \circ S_l = S_l \circ T_v$. The movements S_l and T_v commute, i.e., the two products taken in different order are indeed equal, because the figure $AA_1A'A_2$ is always a rectangle.

Glide reflections, like all other types of plane movements, can be successfully used for solving geometric problems.

Problem 23. *Construct a line parallel to the side AC of a given triangle ABC and intersecting its sides AB and BC at points D and E such that $AD = BE$.*

Solution. The solution relies on the following two properties of glide reflections, which immediately follow from Figure 19:

(1) the midpoint of a segment joining an arbitrary point with its image under a glide reflection always lies on the axis;

(2) the axis of the glide reflection is preserved.

There is a glide reflection U which takes the half-line AB into the half-line BC. Its axis is the line NK, where N is the midpoint of the segment AB while K belongs to BC and $BK = NB$. By the premises, $AD = BE$; hence $U(D) = E$ and the midpoint of DE must belong to the line NK. But, since $DE \parallel AC$, the midpoint of DE lies on the median BM. Therefore the three segments DE, BM and NK have a point in common, and the required construction can be effectuated in the following order. First we find the points N and K as mentioned above. Then we draw the median BM. Finally, we draw the line parallel to AC through the intersection point of BM and NK. This is the desired line.

Exercise 44. A point and three straight lines are given. Draw a line l passing through the given point in such a way that its image under the three reflections with respect to the three given lines (in a prescribed order) is parallel to l.

Exercise 45. Using complex numbers, find an algebraic formula for glide reflection.

7. Classification of movements

In the previous section, we have gotten acquainted with a new kind of plane movement. So far, we have encountered four types of plane movements: translations, rotations, reflections and glide reflections. A natural question arises: *are there any plane movements that do not belong to any of these four types?* The answer is given by the following theorem.

Theorem 4. *Any plane movement is either a translation, a rotation, a reflection, or a glide reflection.*

Proof. First of all, we note that a plane movement is completely defined by the images of three non-collinear points A, B, C. In fact, if A', B', C' are the images of these points, then for any point D there exists exactly one point D' whose distances from A', B', C' are equal to the distances of D from A, B, C.

The second useful observation is that for any two different points M and M' there is a reflection that carries M over to M'. In fact, this reflection is uniquely defined: its axis is the perpendicular bisector of the segment MM'.

Using these two observations, we are going to decompose any plane movement as the product of several reflections. Note that we have already used this trick earlier: see the discussion of Problem 21.

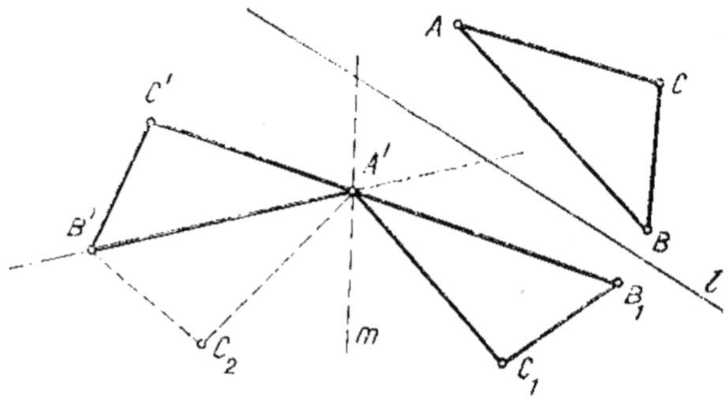

Figure 20. Decomposition of a plane movement into reflections

Let f be an arbitrary movement of the plane. Choose three non-collinear points A, B and C. Denote $f(A) = A'$, $f(B) = B'$, $f(C) = C'$. Suppose that A' is different from A, denote by S_l the reflection that takes A to A', and set $B_1 = S_l(B)$, $C_1 = S_l(C)$ (see Figure 20). If B_1 is different from B', then we denote by S_m the reflection that takes B_1 to B' while preserving A', and set $C_2 = S_m(C_1)$. Finally, if $C_2 \neq C'$, we find a third reflection, S_n, which takes C_2 into C'. We thus see that in the worst case, when all the steps of this procedure are necessary, f can be represented as the composition $S_n \circ S_m \circ S_l$. If some steps turn out to be unnecessary, we can represent f as one reflection or a composition of two reflections.

Now we will prove that the product of no more than three reflections is a movement belonging to one of the four types that we know. Indeed, one reflection is a reflection, and that's it. Two reflections make either a rotation or a translation. The only nontrivial case is to analyze the product of three reflections $S_n \circ S_m \circ S_l$.

Three lines in a plane can be arranged in one of the four essentially different patterns depicted in Figure 21. We will show that in cases (a) and (b), the product is a reflection, and in cases (c) and (d), a glide reflection.

In case (a), the composition $S_m \circ S_l$ of the last two reflections is a rotation through an angle equal to twice the angle between the lines

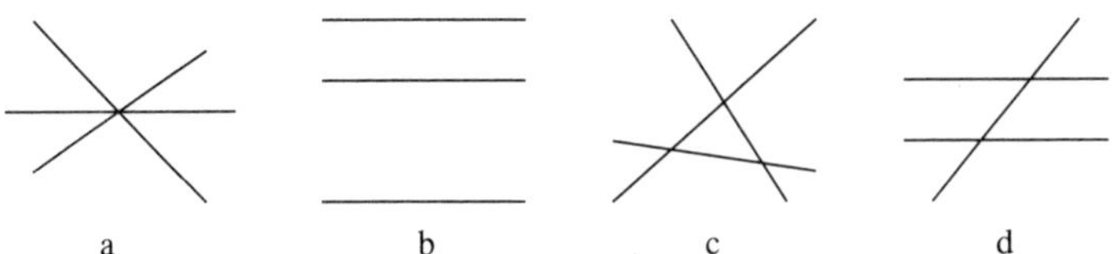

Figure 21. Three lines in the plane

m and l. We can choose another line l' passing through the same point, so that $S_m \circ S_l = S_n \circ S_{l'}$. Then

$$S_n \circ S_m \circ S_l = S_n \circ S_n \circ S_{l'} = S_{l'}.$$

In case (b), a similar argument holds.

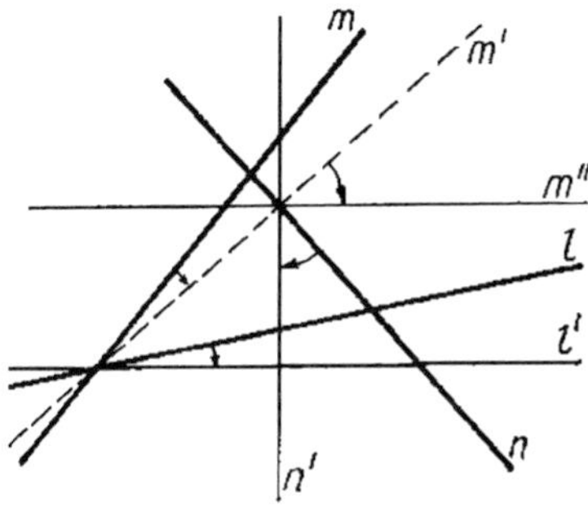

Figure 22. Adjusting two rotations

Now consider case (c). In the initial product of three reflections, we will make two changes. First we replace the product $S_m \circ S_l$ by an equal product $S_{m'} \circ S_{l'}$, where the line m' is chosen to be perpendicular to n (Figure 22). We have $S_n \circ S_m \circ S_l = S_n \circ S_{m'} \circ S_{l'}$. Next we replace the product $S_n \circ S_{m'}$ by $S_{n'} \circ S_{m''}$, where n' is perpendicular to l'. We obtain

$$f = S_n \circ S_m \circ S_l = S_n \circ S_{m'} \circ S_{l'} = S_{n'} \circ S_{m''} \circ S_{l'}.$$

Note that the lines l' and m'' are parallel; therefore the composition $S_{m''} \circ S_{l'}$ is a parallel translation in the direction of line n', and the whole movement is a glide reflection.

Finally, case (d) is reduced to case (c), because the composition $S_n \circ S_m \circ S_l$ remains the same if two of the three lines (n and m or m and l) get rotated by the same angle. $\square$

Plane movements have a simple description in terms of complex functions. Theorem 3, proved above, says that translations and rotations correspond to functions $\alpha z + m$ with $|\alpha| = 1$.

Theorem 5. *The set of reflections and glide reflections of the plane coincides with the set of all transformations described by complex formulas*

$$w = \alpha \bar{z} + m, \tag{18}$$

where α and m are complex numbers and $|\alpha| = 1$.

Proof. The fact that reflections and glide reflections are indeed described by such formulas follows directly from the result of Exercises 37 and 45.

To prove the second half of the theorem, note that the composition of transformation (18) with the standard reflection $z \mapsto \bar{z}$ is given by the formula $z \mapsto \alpha z + m$, which, by Theorem 3, is either a translation or a rotation. $\square$

8. Orientation

We have learned that there are four types of plane movements: translations, rotations, reflections, and glide reflections. Movements of the first two types can be represented as the product of an even number (two) of reflections; they are referred to as *proper* movements. The remaining two types are products of an odd number (one or three) of reflections; they are referred to as *improper* movements, because one has to exit the plane in order to physically implement such a movement.

The distinction between the two kinds of plane movements can be best understood using the notion of *orientation*.

We say that the ordered triple of non-collinear points A, B, C is *positively oriented*, if this ordering agrees with a counterclockwise walk around the triangle ABC, or, in other words, if in the sequence $\overrightarrow{AB}$, $\overrightarrow{BC}$, $\overrightarrow{CA}$ every next vector is a turn to the left with respect to the previous one. If the order is clockwise, the triple is said to be *negatively oriented*.

Figure 23. Three pucks

Exercise 46. Three pucks form a triangle in the plane. A hockey-player chooses a puck and sends it along a straight line so that it passes between the two remaining pucks. Is it possible that after 25 shots each of the three pucks returns to its initial position?

It is remarkable that any movement f of the plane either preserves or reverses the orientation of all triples: the orientation of $f(A)$, $f(B)$, $f(C)$ either coincides with that of A, B, C for all triples, or differs from it for all triples. More specifically, it is easy to see that proper movements (translations and rotations) preserve the orientation, while improper movements (reflections and glide reflections) reverse it.

As a consequence of this observation, we obtain the following fact: the composition of an odd number of reflections can never be an identity transformation.

The notion of orientation has a simple interpretation in terms of complex numbers.

Exercise 47. Prove that the triple (z_1, z_2, z_3) is positively oriented if and only if the argument of the complex number $(z_3 - z_1)/(z_2 - z_1)$ is between 0 and 180°.

9. Calculus of involutions

Definition 9. A transformation f is called an *involution*, if it is not the identity, but its square is the identity: $f \neq \text{id}$, $f^2 = f \circ f = \text{id}$. This is the same as to say that f is inverse to itself: $f = f^{-1}$, that is, $f(A) = B$ if and only if $f(B) = A$.

There are two types of involutive movements of the plane:

- R_A — half turn around point A (see Section 3).
- S_l — reflection in a line l (see Section 2),

We see that involutive movements correspond to geometric elements of two kinds: points and lines. This correspondence is in fact one-to-one, because different points and different lines produce different involutions. Therefore, the passage from geometric objects to involutions preserves all information, and every fact about points and lines can be reformulated in terms of the corresponding involutions.

Problem 24. *Find the property of a pair of reflections S_l, S_m which is equivalent to the fact that the lines l and m are mutually perpendicular.*

Solution. The composition $S_m \circ S_l$ is a translation if $m \parallel l$, or a rotation through 2φ if m and l intersect at an angle φ. Unless $m = l$, this composition can never be the identity. Its square $(S_m \circ S_l)^2$ is either a translation (in the first case) or a rotation through 4φ (in the second case). Hence the lines m and l are perpendicular if and only if

$$(S_m \circ S_l)^2 = \text{id}, \tag{19}$$

i.e., the product $S_m \circ S_l$ is an involution. Note that this involution is a half turn around the intersection point of the two given lines.

If we multiply (19) by S_m on the left and by S_l on the right, then it becomes

$$S_m \circ S_l = S_l \circ S_m, \tag{20}$$

i.e., the two involutions S_m and S_l *commute*. This is the required condition for the two lines to be perpendicular.

This is an appropriate moment to discuss the notions of commutativity and associativity. Multiplication of movements is in general non-commutative. As we have just seen, two different reflections commute if and only if the corresponding lines are perpendicular. But the composition of movements, like that of any arbitrary transformations, always has the property of *associativity.*

Let us be given four sets and three mappings between them, arranged according to the scheme

$$V \xrightarrow{f} W \xrightarrow{g} X \xrightarrow{h} Y.$$

Then one can form the following compositions: $g \circ f : V \to X$, $h \circ g : W \to Y$, $h \circ (g \circ f) : V \to Y$, $(h \circ g) \circ f : V \to Y$. Associativity means that the two double compositions $h \circ (g \circ f)$ and $(h \circ g) \circ f$ coincide.

To find a formal proof of this almost evident property, it is enough to understand the meaning of composition. Thus, the mapping $g \circ f$ is defined by the equation $(g \circ f)(v) = g(f(v))$ for an arbitrary element $v \in V$. In the following chain of equations this definition is used several times:

$$\begin{aligned} (h \circ (g \circ f))(v) &= h((g \circ f)(v)) = h(g(f(v))) \\ &= (h \circ g)(f(v)) = ((h \circ g) \circ f)(v). \end{aligned} \tag{21}$$

Since the values of $h \circ (g \circ f)$ and $(h \circ g) \circ f$ on any element are the same, these two mappings coincide.

The following analogy might be useful to better understand the meaning of associativity. Imagine that f, g and h are the actions of putting on your socks, boots and overshoes, respectively. Then the composition $(h \circ g) \circ f$ means that one first puts on the socks, then puts the boots inside the overshoes, and puts this object on the feet in socks. The other composition $h \circ (g \circ f)$ means that one first puts the socks inside of the boots, puts on this combination and then puts the overshoes on top. Evidently, the result in both cases is the same!

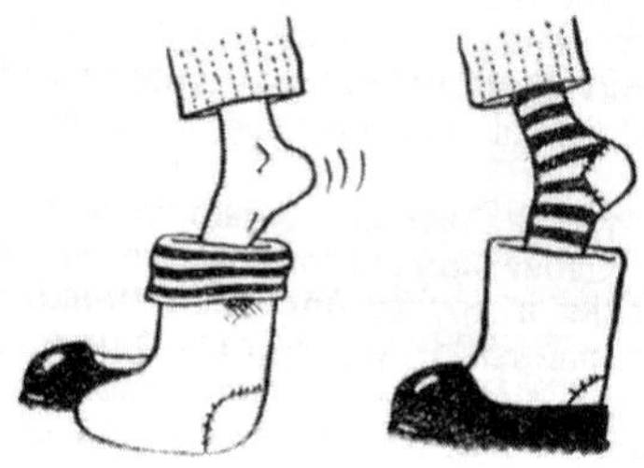

Figure 24. Associativity

The same analogy shows that the composition of operations in question is not commutative: to put on the socks, then the boots is not the same thing as to put on the boots, then the socks! However, commuting operations do exist, for example, putting a sock on one foot and putting a sock on another foot.

Using this analogy, it is easy to understand the formula for the operation inverse to a composition of several operations. For example, if you put on the socks, then the boots, then the overshoes, then the inverse operation means that you take off first the overshoes, then the boots, then the socks:

$$(h \circ g \circ f)^{-1} = f^{-1} \circ g^{-1} \circ h^{-1}.$$

Now we return to the calculus of involutions in the plane.

Problem 25. *Express in terms of involutions the property of four points A, B, C, D forming a parallelogram.*

Solution. According to Exercise 42, the composition $R_A \circ R_B$ is translation by the vector $2\,\overrightarrow{BA}$, while the composition $R_D \circ R_C$ is translation by the vector $2\,\overrightarrow{CD}$. The figure $ABCD$ (with this order of vertices!) is a parallelogram if and only if $\overrightarrow{BA} = \overrightarrow{CD}$, which is equivalent to the following condition for the four involutions:

$$R_A \circ R_B = R_D \circ R_C.$$

Multiplying both sides of this relation by appropriate involutions, we can rewrite it in two more equivalent forms: $R_A \circ R_B \circ R_C \circ R_D = \mathrm{id}$ and

$$R_A \circ R_B \circ R_C = R_D. \tag{22}$$

The last equation may be viewed as a formula that expresses the fourth vertex of a parallelogram in terms of the three given ones.

Exercise 48. Express the following geometric facts as algebraic relations between the corresponding involutions: (a) point A belongs to the line l; (b) point A is the midpoint of the segment BC.

Exercise 49. Find the geometric meaning of the following relations: (a) $R_A \circ S_l = S_l \circ R_B$; (b) $(S_n \circ S_m \circ S_l)^2 = \mathrm{id}$.

You can see that the algebra of involutions often provides a short and convenient way to write down facts about points and lines in the plane. Here is a more complicated example where this technique is essential.

Problem 26. *Let M, N, P, Q be the centres of the four squares built on the sides of a quadrangle $ABCD$ (Figure 25). What conditions should be imposed on $ABCD$ in order that $MNPQ$ be a square?*

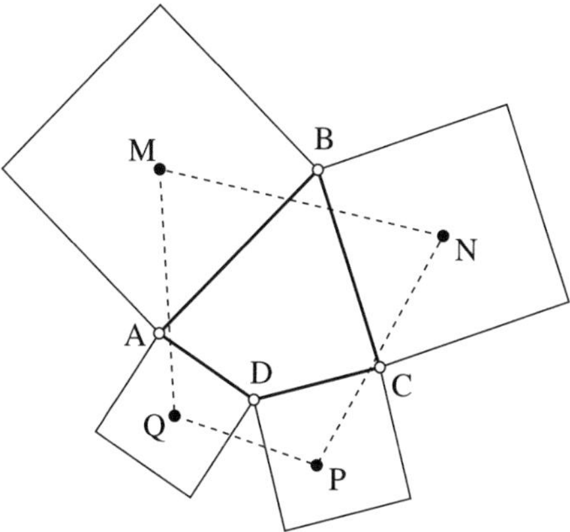

Figure 25. Squares on the sides of a quadrilateral

Solution. We know that the diagonals of $MNPQ$ are always equal and mutually perpendicular (see Exercise 41). Therefore, $MNPQ$ is a square if and only if it is a parallelogram. Using the result of Problem 25, we can write this as the following condition on the four involutions:

(23) $$R_M \circ R_N \circ R_P \circ R_Q = \mathrm{id}\,.$$

By formula (14) we have

$$\begin{aligned} R_M &= R_A^d \circ R_B^d, \\ R_N &= R_B^d \circ R_C^d, \\ R_P &= R_C^d \circ R_D^d, \\ R_Q &= R_D^d \circ R_A^d, \end{aligned}$$

where $d = 90°$, and we recall that if the angle of rotation is not specified, it is assumed to be 180°. Upon substitution into (23), this gives

(24) $$R_A^d \circ R_B \circ R_C \circ R_D \circ R_A^d = \mathrm{id},$$

or, after multiplication by R_A^{-d} on both sides,

$$R_B \circ R_C \circ R_D = R_A.$$

According to formula (22), this means that $ABCD$ is a parallelogram. Hence, the necessary and sufficient condition for $MNPQ$ to form a square is that the initial quadrangle $ABCD$ be a parallelogram.

Chapter 3

Transformation Groups

The notion of a group unifies two different ideas: a geometric one and an algebraic one.

On the geometric side, the notion of a *transformation group* gives a mathematical expression of the general principle of symmetry: the more transformations preserve a given object, the more symmetric it is.

On the algebraic side, the notion of an *abstract group* contains the common features of operations that most often appear in mathematics. Examples of such operations — addition and multiplication of numbers and points, addition of vectors, composition of movements — were considered in the previous chapters.

1. A rolling triangle

We begin with an introductory problem where a transformation group comes up in a natural way.

Problem 27. *An equilateral triangle ABC lies on the plane. One can roll it over the plane by turning it through* $180°$ *around any of its sides. Show that if after a certain number of such steps the triangle returns to the initial place, then each of its three vertices will return to its initial position.*

Solution. Let a, b and c be the lines containing the sides of the given triangle in its initial position. After any number of turnovers the sides of the triangle will lie

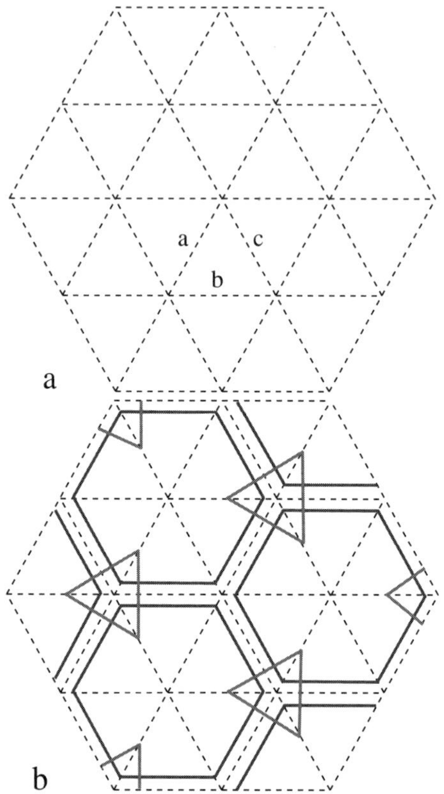

Figure 1. Ornament of the rolling triangle

on the lines of the triangular network shown in Figure 1a.

The allowed transformations are compositions of reflections in these lines. Let G be the set of all such transformations. This set contains, for example, rotations through 120° around the vertices, and glide reflections whose axes coincide with the middle lines of our triangle.

The problem will be solved if we prove that the only plane movement that belongs to G and leaves the triangle in its place is the identity transformation. Apart from the identity, there are five movements that take the triangle into itself: two nontrivial rotations around its centre and three reflections in its altitudes. We have to show that none of these belongs to the set G.

The reader has probably encountered problems that are solved by constructing an appropriate example (or counterexample). We will use the same trick here — but the example we are going to construct is unusual: it is an *ornament* possessing the following two properties. First, it is symmetric with respect to any of the lines shown in Figure 1a. Second, it is *not* symmetric with respect to the altitudes of the triangle ABC, and its centre is *not* the centre of rotational symmetry for the ornament. The first property implies that the ornament is preserved by any movement that belongs to the set G, while the second means that G contains none of the five nontrivial movements of the triangle. Thus we prove the required result.

It remains to construct an ornament with all the specified properties. An appropriate example is provided by an ancient Chinese ornament (a grating) shown in Figure 1b. Of course, this example is not unique. The general recipe to build such an example can be stated as follows. Choose a completely asymmetric pattern inside the triangle ABC, i.e. a figure which is not preserved by any non-identity movement of the triangle. Then take the union of all figures that are obtained from this pattern by successive reflections in the sides of the triangle. One may imagine that the pattern inside the triangle is dyed with paint, leaving a colour trace on the plane when the triangle rolls over.

Exercise 50. Introduce a coordinate system in the plane (see definition 3 on page 20) such that the point A has coordinates $(0,0)$, point B has coordinates $(6,0)$, while point C has coordinates $(0,6)$

(triangle ABC is still supposed to be equilateral!). Take the pattern consisting of one point $K(3,1)$. Draw the ornament resulting from this pattern, and describe the coordinates of all its points.

Exercise 51. Will the assertion of Problem 27 still hold if the equilateral triangle is replaced by a triangle with angles (a) 45°, 45°, 90°? (b) 30°, 60°, 90°? (c) 30°, 30°, 120°?

2. Transformation groups

Definition 10. A *transformation group* is a set G of transformations of a certain set which has the following two properties:

(1) If two transformations f and g belong to G, so does their composition $f \circ g$.

(2) Together with every transformation f, the set G also contains the inverse transformation f^{-1}.

These two properties mean that the elements of the set G are interrelated and form a whole which is closed under composition and taking the inverse. The notion of a set closed with respect to a certain operation has appeared several times in this book, starting from Problem 1.

Example 1. The set of transformations G considered in the discussion of Problem 27 is a transformation group. Property (1) is crucial for the solution of Problem 27. It holds by construction. Property (2) is also valid because the inverse to a series of reflections is the series of the same reflections performed in the inverse order.

Example 2. The set of all transformations of a given set M forms a transformation group, denoted by $\mathrm{Tr}(M)$.

Exercise 52. Prove that every transformation group contains the identity transformation.

Apart from the group G, in the discussion of Problem 27 we have dealt with two more groups: the group $\mathcal{M}$ of all movements of the plane and the symmetry group D_3 of symmetries of an equilateral triangle.

The groups G and D_3 are contained in $\mathcal{M}$; this fact is usually expressed by saying that they are *subgroups* of $\mathcal{M}$. In general, given an arbitrary plane figure Φ, one can consider the set of all plane movements that take Φ into itself. This set is denoted by $\mathrm{Sym}(\Phi)$ and called the *symmetry group* or *group of movements* of the figure Φ. Every element of this group is called a *symmetry* of the figure Φ. Thus, D_3 is the symmetry group of an equilateral triangle: $\mathrm{Sym}(\triangle) = D_3$. The group G in Problem 27 is also a symmetry group of a certain figure, namely, of the ornament depicted in Figure 1b. To verify this fact, we only have to check that any movement preserving the ornament is a composition of several reflections in the lines of the triangular grid.

Exercise 53. Make sure this is indeed true.

The notion of a symmetry group is a source of numerous interesting examples of transformation groups. Let us consider some of them.

Problem 28. *Figure 2 shows ancient Japanese family insignias (kamon). Find the symmetry group of each kamon. Which symmetry groups are the same and which are different?*

Figure 2. First set of *kamon*

Solution. Figure Φ_1 is axially symmetric with respect to the four axes at angles of 45° from each other; it also does not change under rotations through 90°, 180° and 270°. The group $\mathrm{Sym}(\Phi_1)$, like any transformation group, also contains the identity transformation; therefore, the total number of its elements is 8. The group $\mathrm{Sym}(\Phi_2)$ also contains eight elements, but it is different from $\mathrm{Sym}(\Phi_1)$, because all its elements are rotations. Figure Φ_3 has the same symmetry as Φ_1: four reflections and four rotations, including the identity. Finally, figure Φ_4 has no nontrivial transformations, and its symmetry group consists of only one element, the identity transformation.

Exercise 54. Find the symmetry groups of Figures 3, a—d. Compare them with each other, and also with the groups of Problem 28.

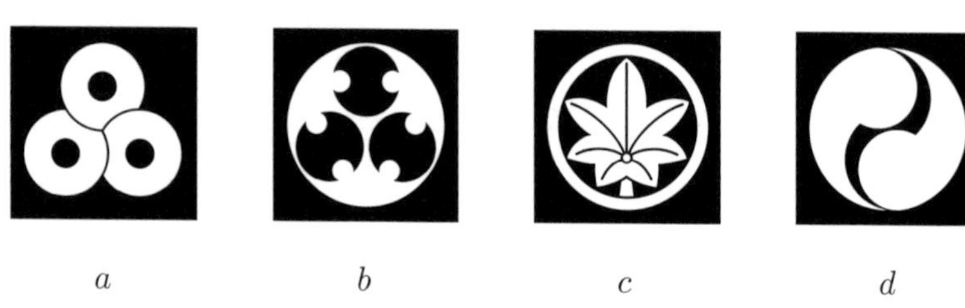

a *b* *c* *d*

Figure 3. Second set of *kamon*

3. Classification of finite groups of movements

The symmetry groups of all the kamon displayed in Figures 2 and 3 consist either of several rotations or of several rotations and an equal number of reflections (including the identity). This observation can be generalized as a theorem. To state it, we need some terminology.

A transformation group is said to be *finite* if it consists of a finite number of elements. The symmetry groups of all the kamon in Figures 2 and 3 are finite. A group which which consists of an infinite number of elements is called an *infinite group*. The group $\mathcal{M}$ of all plane movements and the group G (Problem 27) are infinite.

The *order* of a group is the number of elements it contains. A finite group is a group of finite order.

The group that consists of rotations around a common centre through multiples of $360°/n$ is called the *cyclic group of order* n and denoted by C_n. The group that contains the same rotations and n reflections in the lines passing through the same centre and such that the angle between any two neighbouring lines is $180°/n$ is called the *dihedral group of order* $2n$ and denoted by D_n.

For example, the symmetry groups of the eight figures that we considered (Figures 2 and 3) are D_4, C_8, D_4, C_1, C_3, D_3, D_1, C_2.

Theorem 6. *Any finite group of plane movements is either a* C_n *or a* D_n.

Proof. To prove the theorem, we first note that a finite group cannot contain parallel translations, because, if it contains a translation by vector $\mathbf{a}$, it must also contain an infinite number of translations by multiple vectors $n\mathbf{a}$.

If the group contains a glide reflection, it also contains its square, which is a parallel translation, and therefore cannot be finite. We conclude that any finite group of plane movements consists entirely of rotations and reflections.

All rotations belonging to the group must have a common centre, because the following exercise shows that a group containing two rotations with different centres also contains a parallel translation.

Exercise 55. Prove that if A and B are two different points of the plane and the angles φ and ψ are not multiples of $360°$, then the product $R_B^{-\psi} \circ R_A^{-\varphi} \circ R_B^{\psi} \circ R_A^{\varphi}$ is a nontrivial translation.

Denote all rotations that belong to a given finite group by R_A^0, R_A^{φ}, ..., R_A^{ω}, where the angles $\varphi, \ldots, \omega$ are chosen to be positive and not exceeding $360°$. Suppose that ψ is the smallest of these angles. Then all the remaining angles must be multiples of ψ. Indeed, suppose that φ is not divisible by ψ. Then it can be written as $\varphi = k\psi + \xi$, where k is an integer and $0 < \xi < \psi$. The rotation through ξ must also belong to the group under study, and we arrive at a contradiction.

Note that ψ must be equal to $360°/n$ for some integer n—otherwise a certain power of the rotation R_A^{ψ} would represent a rotation by an angle smaller than ψ. We have thus proved that all the rotations present in any finite group of plane movements are $R_A^{k\psi}$, where $k = 0, 1, \ldots, n-1$ and $\psi = 360°/n$. If the group contains no reflections, then it is the group C_n.

Now suppose that the group contains n rotations and at least one reflection. We are going to prove that the number of reflections in the group is exactly n.

Indeed, if R_1, R_2, $\ldots$, R_n are n different rotations and S a reflection, then the n compositions $S \circ R_1$, $S \circ R_2$, $\ldots$, $S \circ R_n$ represent n different reflections that belong to the group. Thus, the number of reflections is no smaller than n. Similarly, the number of rotations in the group is no smaller than the number of reflections, because, if S_1, S_2, $\ldots$, S_m are m different reflections, then $S_1 \circ S_1$, $S_1 \circ S_2$, $\ldots$, $S_1 \circ S_m$ are m different rotations (including the identity).

We have thus proved that any finite group of plane movements either is a C_n or it consists of n rotations with a common centre and an equal number of reflections. If $n = 1$, what we get is the group D_1 of order 2, which contains one reflection and the identity transformation. If $n \geq 2$, we have to show that the axes of all reflections pass through the centre of rotations. First, observe that a finite group may not contain two reflections whose axes are parallel, because their product would produce a translation by a non-zero vector. Thus any two axes must have a common point. The product of the two reflections whose axes intersect at a point P making the angle φ belongs to the group and is a rotation around P through the angle 2φ. Hence the point P coincides with the common centre A of all rotations, while the angle φ is a multiple of $180°/n$. The group under study is thus D_n. This completes the proof. □

Exercise 56. Can a plane figure have

(1) exactly two symmetry axes?

(2) exactly two centres of symmetry?

Exercise 57. Which is the most symmetrical (i.e., having the biggest symmetry group) bounded plane figure?

4. Conjugate transformations

In the discussion of Problem 28 above we said that the symmetry groups of figures Φ_1 and Φ_3 are the same: $\mathrm{Sym}(\Phi_1) = \mathrm{Sym}(\Phi_3)$. What is the precise meaning of this equality? In a more general setting: what is the precise meaning of the classification theorem we proved in the previous section? Let us think a little about these questions.

Equality of two symmetry groups like Sym(Φ_1) and Sym(Φ_3) has the verbal meaning, i.e., the two sets of plane movements coincide only if the two figure are placed in the plane in such a way that their centres and symmetry axes coincide. Otherwise the sets $G = \text{Sym}(\Phi_1)$ and $H = \text{Sym}(\Phi_3)$ would be different, although closely related to each other. We will now elucidate this relation.

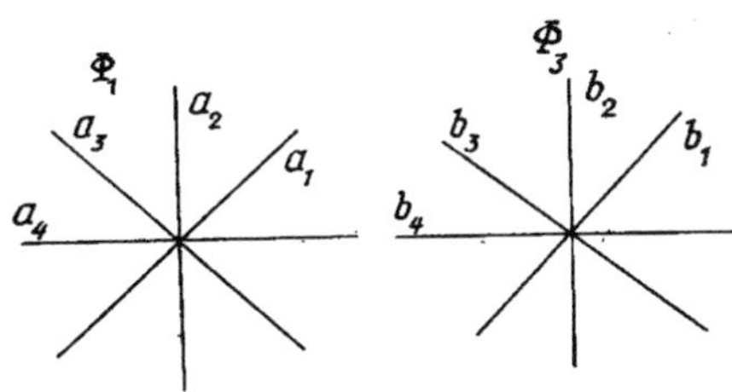

Figure 4. Conjugate symmetry groups

Denote the centre of rotations and the symmetry axes for the group G by A and $a_1, \ldots, a_4$. Denote the same objects for the group H by B and $b_1, \ldots, b_4$, respectively (see Figure 4). Let f be a plane movement that carries B into A and each b_i into a_i. The transformations of the group G can be obtained from the transformations that belong to the group H in the following way. Consider, for example, the reflection in the line b_3. First move the figure Φ_1 with the help of the movement f^{-1}, then apply reflection in b_3 to it, then move it back with the help of f. It is easy to see that after all these actions the figure has undergone a reflection in the line a_3.

In general, for any element h of the group H the composition $f \circ h \circ f^{-1}$ is an element of G. More precisely, if h is a reflection in b_i, then $f \circ h \circ f^{-1}$ is a reflection in a_i; if h is a rotation around B, then $f \circ h \circ f^{-1}$ is a rotation around A through the same angle.

Definition 11. Two transformations $h \in H$ and $g = f \circ h \circ f^{-1} \in G$ are said to be *conjugate*. The transition from h to g is called *conjugation by f*.

We have encountered conjugate transformations before, when we derived the complex number formula for a rotation with an arbitrary centre, and in Exercise 37 (Chapter 2).

Two groups of movements are said to be *conjugate*, if the list of elements of one group becomes the list of elements of the other upon conjugation by a certain (one and the same) movement. The symmetry groups of two copies of one and the same figure, placed arbitrarily in the plane, are always conjugate. The conjugation is effectuated by a movement that carries one of the copies into the other.

In general, conjugation should be understood as looking at an object from a different viewpoint. The conjugating movement is the one which relates the two viewpoints (or systems of reference, in physical terminology).

The most important property of conjugate subgroups is that they have the same internal structure. Let us explain the exact mathematical meaning of this phrase. Let G and H be two groups of plane movements that are conjugated by a movement f. If $g = f \circ h \circ f^{-1}$, then we will say that g and h *correspond* to each other, and write $g \leftrightarrow h$. This correspondence is one-to-one, because h can be uniquely expressed in terms of g as $f^{-1} \circ g \circ f$.

Then the following two facts hold:

(1) If $g_1 \leftrightarrow h_1$ and $g_2 \leftrightarrow h_2$, then $g_1 \circ g_2 \leftrightarrow h_1 \circ h_2$.

(2) If $g \leftrightarrow h$, then $g^{-1} \leftrightarrow h^{-1}$.

Both facts are verified in a straightforward way:

(1) $g_1 \circ g_2 = (f \circ h_1 \circ f^{-1}) \circ (f \circ h_2 \circ f^{-1}) = f \circ (h_1 \circ h_2) \circ f^{-1}$.

(2) $(f \circ h \circ f^{-1})^{-1} = f \circ h^{-1} \circ f^{-1}$.

Thus, both group operations (composition and taking the inverse) in one group correspond to their respective counterparts in another group under the correspondence under study. This phenomenon is called *isomorphism*, and we will study it in detail later in this chapter.

Now we will derive formulas for conjugation in the group of plane movements.

Problem 29. *Find the movement which is conjugate to the rotation* R_A^α *by means of the reflection* S_l.

Solution. By virtue of the general remark that we made above, to get the conjugate movement one has to look at the given rotation 'from under the plane', displacing one's viewpoint by means of the reflection S_l. It is fairly evident that the result is the rotation through $-\alpha$ around the point A' which is symmetric to A with respect to l. We will perform a rigorous check of this result, i.e. prove that $S_l \circ R_A^\alpha \circ S_l^{-1} = R_{A'}^{-\alpha}$.

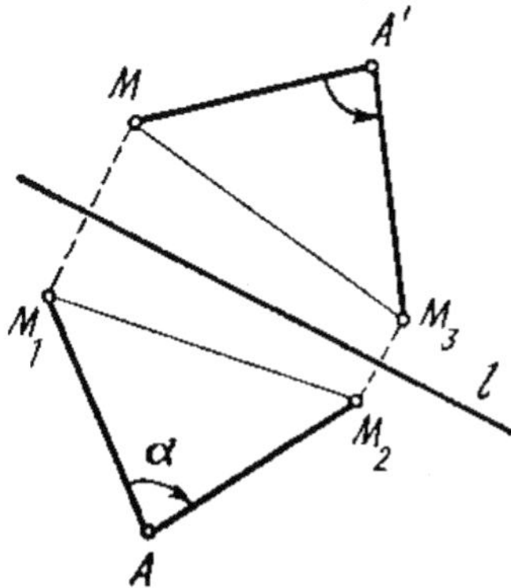

Figure 5. Conjugation of a rotation by a reflection

Indeed, let M be an arbitrary point in the plane (see Figure 5). Let M_1 be its image under the movement S_l^{-1} (which, in fact, is the same thing as S_l). Suppose that M_1 goes into M_2 under R_A^α and M_2 goes into M_3 under S_l. Then $\triangle MA'M_3 = \triangle M_1AM_2$, whence $M_3 = R_{A'}^{-\alpha}(M)$.

Exercise 58. Compile a complete table of all conjugations in the group of plane movements, i.e., find $f \circ g \circ f^{-1}$ for every $g = T_{\mathbf{a}}, R_A^\alpha, S_l, U_l^{\mathbf{a}}$ and every $f = T_{\mathbf{b}}, R_B^\beta, S_m, U_m^{\mathbf{b}}$.

To acquire freedom in manipulations with plane movements, we suggest that the reader make and practice using a simple tool that we call a 'dihedral instrument'. It consists of a regular polygon cut from cardboard and of the same polygon drawn on a piece of paper. The vertices of each polygon should be consecutively numbered by 1, 2, ...

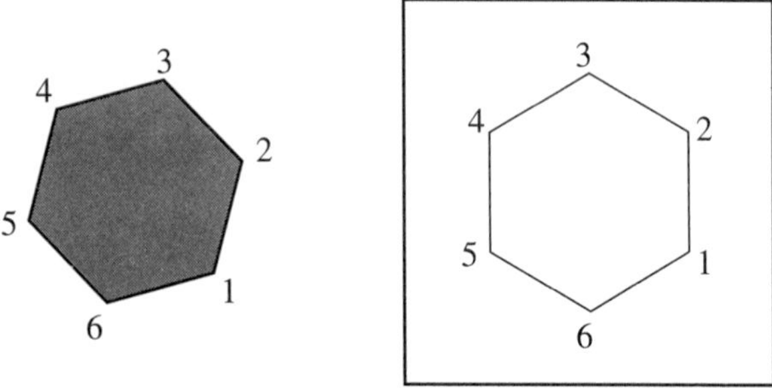

Figure 6. Instrument for studying the dihedral group

With the help of this instrument, one can study the dihedral group D_n. To find the product of two elements of D_n, one must first place the cardboard polygon on the paper one in its initial position, so that the vertices with the same numbers coincide, then perform the given movements one after another and, comparing the numbers of vertices, try to figure out what is the composed movement. In the same way one can also compute the table of conjugations.

Exercise 59. Using appropriate dihedral instruments, fill in the multiplication and conjugation tables for the groups D_3 and D_4.

Looking at the table of conjugate movements (the answer to Exercise 58), one can see that a translation $T_{\mathbf{a}}$ does not change when conjugated by another translation $T_{\mathbf{b}}$:

$$T_{\mathbf{b}} \circ T_{\mathbf{a}} \circ T_{\mathbf{b}}^{-1} = T_{\mathbf{a}}.$$

This equation is equivalent to $T_{\mathbf{b}} \circ T_{\mathbf{a}} = T_{\mathbf{a}} \circ T_{\mathbf{b}}$, which means that any two translations *commute*; in other words, they are interchangeable with respect to composition. (We recall that we have already mentioned commutativity in the discussion of Problem 24.)

A transformation group where any two elements commute is called *commutative*.

Problem 30. *Find all finite commutative groups of plane movements.*

Solution. We know the list of all finite groups of plane movements: it is made up of C_n and D_n for all $n = 1, 2, 3, \ldots$.

Every cyclic group C_n is commutative. This follows from the fact that the group of all rotations with a common centre is commutative, because the product of rotations through angles α and β is a rotation through $\alpha + \beta$, no matter which order the composition is taken in.

The group D_1 is commutative, because it consists of only two elements, one of which is the identity. It is easy to see that the group D_2 is commutative, too. In fact, it consists of two reflections S_1 and S_2 whose axes are mutually perpendicular, a half-turn rotation R and an identity transformation. According to the rules we have derived earlier, the composition $S_1 \circ S_2$ is a rotation through 180° in one direction, while $S_2 \circ S_1$ is a rotation through 180° in the opposite direction. Hence, $S_1 \circ S_2 = S_2 \circ S_1$. Therefore, $S_1 \circ R = S_1 \circ S_2 \circ S_1 = R \circ S_1$. Similarly, $S_2 \circ R = R \circ S_2$.

Now consider the group D_n for $n \geq 3$. Choose two reflections that belong to this group and whose axes are adjacent, i.e. make an angle of $180°/n$. Their products are rotations through the angle $360°/n$, in the positive or negative direction, depending on the order in which the product is taken. Since $n \geq 3$, these two products are different. Therefore, the group D_n is non-commutative.

The complete list of finite commutative groups of plane movements thus includes D_1, D_2 and all groups C_n.

The property of commutativity of a group can be easily read off its multiplication table: a group is commutative if and only if its multiplication table is symmetric with respect to the main diagonal. Readers who did Exercise 59 have visual evidence of the fact that the groups D_3 and D_4 are not commutative. However, even if the table is not symmetric as a whole, it always contains pairs of equal elements that occupy symmetric positions. Such pairs correspond to the pairs of commuting elements of the group.

Exercise 60. Indicate all pairs of commuting elements in the groups D_3 and D_4.

5. Cyclic groups

There is another way to explain why the cyclic group C_n is commutative. Let R denote the rotation through $360°/n$. Then all the elements of the group can be represented as powers of R, i.e. $R^2 = R \circ R$, $R^3 = R \circ R \circ R$, $\ldots$, $R^n = \mathrm{id}$, and it is clear that we always have $R^k \circ R^l = R^{k+l} = R^l \circ R^k$.

A transformation f whose powers exhaust the set of all elements of a group is called a *generator*, or a *generating element* of the group. To better understand the meaning of this notion, let us imagine that we have no group, but only one transformation f of a certain set M. The question is whether there exists a transformation group that contains this transformation f. The answer to this question is always positive. The smallest group containing a given transformation f can be constructed in the following way.

If a group contains an element f, then, according to the first defining property of a group, it must contain all its powers f^2, f^3, etc. By the second property, it also must contain the inverse transformation f^{-1} and therefore all its powers $(f^{-1})^2$, $(f^{-1})^3$, etc.

Exercise 61. Prove that $(f^{-1})^k = (f^k)^{-1}$.

The transformation $(f^{-1})^k$, where k is a natural number, is called a negative power of f and is also denoted by f^{-k}. The zeroth power of any transformation is by definition the identity transformation. Now observe that the set of all integer powers of a given transformation, $\ldots$, f^{-2}, f^{-1}, f^0, f^1, f^2, $\ldots$, *always* forms a group, because of the identities $f^k \circ f^l = f^{k+l}$ and $(f^k)^{-1} = f^{-k}$, which hold not only for natural, but also for all integer values of k and l. This set of all powers of f is called *the group generated by* f.

Two possibilities may arise.

(1) All the powers f^k are different. In this case the group generated by f is infinite and is called an *infinite cyclic group.*

(2) Among the powers of f there are some that coincide. Then there is a positive power of f which is equal to the identity transformation. Indeed, if f^k and f^l is any pair of coinciding powers with $k > l$, then $f^{k-l} = \text{id}$. Denote by n the smallest positive exponent satisfying $f^n = \text{id}$. The number n is called the *order* of the transformation f. In this case the group generated by f consists of exactly n different transformations f, f^2, $\ldots$, f^n. (All of these are different indeed, because if we had $f^k = f^l$ with $0 < l < k \leq n$, then we would get $f^{k-l} = \text{id}$, contrary to the choice of n.)

In this case the group generated by f is a *finite cyclic group of order n.* In particular, this notion includes the cyclic groups of rotations C_n considered above.

When the transformation f generates an infinite group, we can also say that f has *infinite order.* The order of the identical transformation is 1 by definition, and the group it generates consists of only one element and is called the *trivial group.* An involutive transformation generates a group of order 2, consisting of itself and the identity.

Problem 31. *Which elements of the group C_{12} are generators of this group? What are the subgroups generated by other elements?*

Solution. C_{12} consists of 12 rotations through angles which are multiples of 30°. All these rotations are powers of the rotation through 30°, which is therefore a generator of the group. The inverse rotation (by 330°) is obviously a generator, too. To facilitate the study of the other elements, we use Figure 7, where every element of the group is represented by a vertex of the regular 12-gon: A_0 corresponds to the identity, A_1 is the rotation through 30°, etc. Consider the elements of the group one by one and, for every element, mark all the vertices of the

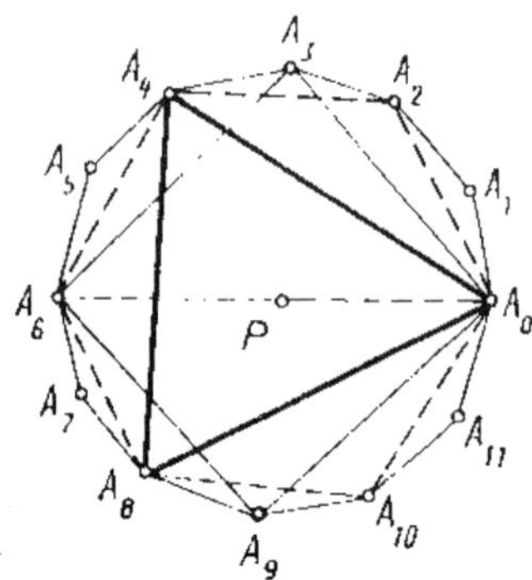

Figure 7. Cyclic group C_{12} and its subgroup

polygon that correspond to the powers of this element. We will see that:

- There are two more rotations — through 150 and 210 degrees — that generate the whole group.
- Rotations through 60 and 300 degrees have order 6 and generate the group C_6 visualized as a hexagon $A_0A_2A_4A_6A_8A_{10}$ in Figure 7.
- Rotations through 90 and 270 degrees generate the group of order 4 (the square $A_0A_3A_6A_9$).
- Rotations through 120 and 240 degrees generate the group of order 3 (the triangle $A_0A_4A_8$).
- The rotation through 180°, which is an involution, generates the group C_2 depicted as the line segment A_0A_6.

This result can be summarized in a table where the upper line is for the values of k, while the lower line shows the order of the rotation R^k (R being the rotation through 30°):

0	1	2	3	4	5	6	7	8	9	10	11
1	12	6	4	3	12	2	12	3	4	6	12

Exercise 62. Using the notion of the greatest common divisor (GCD) of two numbers, find a general formula for the order of the element f^k in a cyclic group of order n generated by f.

A transformation belonging to a finite group is a generator of this group if and only if its order is equal to the order of the group. The number of generating elements in the cyclic group C_n is denoted by $\varphi(n)$, and the function φ is called the *Euler function.* For example, the table above shows that $\varphi(12) = 4$.

Exercise 63. (a) Compile the table of values of the function $\varphi(n)$ for $n = 2, 3, \ldots, 15$. (b) Find a general formula for $\varphi(n)$ in terms of the prime decomposition of the number n.

Now let us see what are the orders of different plane movements according to their type. It is clear that non-identical translations and glide reflections have an infinite order, because under the repeated action of such a transformation any point occupies infinitely many new positions (Figure 8, a and b).

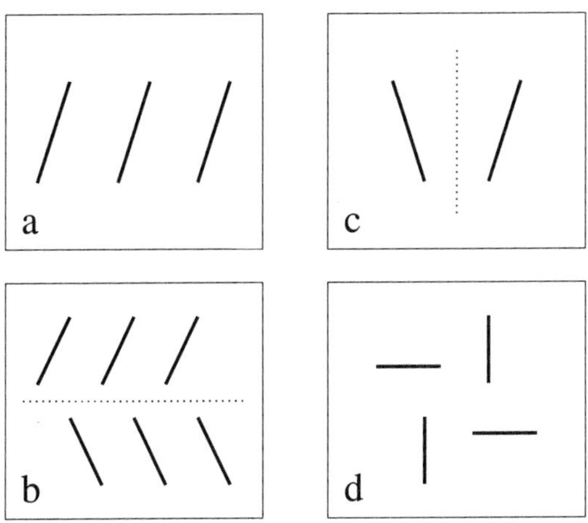

Figure 8. Order of translation, reflection and glide reflection

Exercise 64. Is it possible for a plane figure to remain unmoved under a non-trivial translation or a glide symmetry?

Any reflection has order 2 (Figure 8c).

Rotations may have different orders. If the angle of rotation is measured by a rational number of degrees $360° \cdot m/n$ where m/n is an irreducible fraction, then the rotation has a finite order n (Figure 8d). For an irrational number of degrees the order is infinite.

Exercise 65. Verify the previous assertion for the rotations of Problem 31. Prove the general fact.

6. Generators and relations

Cyclic groups, i.e., groups generated by one element, constitute the simplest class of groups. Now we will consider the groups that cannot be generated by one element.

To begin with, let us prove that the group D_n, where $n \geq 2$, is not cyclic. In fact, we have already seen that if $n > 2$, then D_n is not commutative and hence is not cyclic. In the case $n = 2$, note that every non-identity element of the group D_2 is of order 2, and thus the group does not contain any element of order 4.

A natural question arises: what is the smallest set of elements of the group D_n which generates the whole group, i.e., allows us to express any element of the group using multiplications and taking the inverse? It turns out that two elements are enough, and $\{R, S\}$, where R is the rotation through $360°/n$ and S an arbitrary reflection, is an example of such a set. In fact, every rotation belonging to D_n can be represented as R^k, and every reflection as $R^k \circ S$.

The first of these two assertions is evident. To prove the other, note that all the movements $R^k \circ S$ for $k = 0, 1, \dots, n-1$ are different. Indeed, an equality $R^k \circ S = R^l \circ S$, when multiplied by S on the right, would imply $R^k = R^l$, which is a contradiction. Now observe that all these movements are improper, i.e. reflections, not rotations. Since the total number of reflections in the group D_n is n, we deduce that each of them must appear in the list S, $R \circ S$, $\dots$, $R^{n-1} \circ S$.

We have thus proved that the pair $\{R, S\}$ is a set of generators of the group

$$D_n = \{\mathrm{id}, R, \dots, R^{k-1}, S, R \circ S, \dots, R^{k-1} \circ S\}.$$

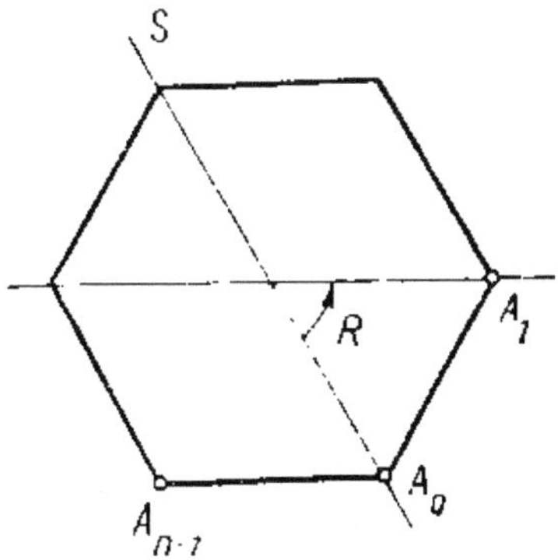

Figure 9. The group D_n

Let us describe the multiplication rule of the group D_n in terms of these expressions through R and S.

- The composition of R^k and R^l is R^{k+l}. If $k+l>n$, this can be replaced by R^{k+l-n} to coincide literally with an element of the above list.
- In the same way, $R^k \circ (R^l \circ S) = R^{k+l} \circ S$, or, if $k+l>n$, it is better to write $R^k \circ (R^l \circ S) = R^{k+l-n} \circ S$.
- Let us find the composition $(R^k \circ S) \circ R^l$. Recall (Problem 29) that $S \circ R^k \circ S = R^{-k}$. Multiplying this equality by S on the right, we get $S \circ R^k = R^{-k} \circ S$. Therefore,

$$(R^k \circ S) \circ R^l = R^k \circ (S \circ R^l) = R^k \circ (R^{-l} \circ S) = R^{k-l} \circ S.$$

- A similar argument shows that $(R^k \circ S) \circ (R^l \circ S) = R^{k-l}$.

In terms of the generators R and S, the multiplication table for the group D_n can be represented as follows:

	R^l	$R^l \circ S$
R^k	R^{k+l}	$R^{k+l} \circ S$
$R^k \circ S$	$R^{k-l} \circ S$	R^{k-l}

where, when necessary, the exponent of R can be increased or decreased by n, using the fact that $R^n = \mathrm{id}$.

For specific values of n, this short table can be expanded to its full form. For example, if $n = 3$, we get the following full table:

	id	R	R^2	S_a	S_b	S_c
id	id	R	R^2	S_a	S_b	S_c
R	R	R^2	id	S_b	S_c	S_a
R^2	R^2	id	R	S_c	S_a	S_b
S_a	S_a	S_c	S_b	id	R^2	R
S_b	S_b	S_a	S_c	R	id	R^2
S_c	S_c	S_b	S_a	R^2	R	id

An important observation is that the complete multiplication table of the group D_n follows from just three relations between the generating elements R and S:

$$S^2 = \mathrm{id}; \quad (S \circ R)^2 = \mathrm{id}; \quad R^n = \mathrm{id}\,. \tag{25}$$

All other relations between R and S can be formally deduced from these three using the definition of a group. As an example, let us check this for the relation $S \circ R^k \circ S = R^{-k}$ that we have used when working on the multiplication table.

The second relation in (25) can be expanded as

$$S \circ R \circ S \circ R = \mathrm{id}$$

or, equivalently, as

$$S \circ R \circ S = R^{-1}.$$

Taking into account that $S^2 = \mathrm{id}$, the n-th power of the last equality gives

$$S \circ R^k \circ S = R^{-k}.$$

It turns out that relations (25) constitute a complete set of *defining relations* for the group D_n in the following sense: if a group is generated by two elements S and R which satisfy the relations (25) and *no other relation except for those that are formal consequences of these three*, then the order of this group is $2n$ and its structure is the same as the structure of the group D_n.

We will now try to give a general definition of generators and defining relations of a transformation group. Let $S = \{s_1, \ldots, s_n\}$ be a finite subset of a group G. By a *monomial over* S we understand a product of the form $M = s_{i_1}^{k_1} s_{i_2}^{k_2} \cdots s_{i_m}^{k_m}$, where i_1, ..., i_m are numbers between 1 and n and the exponents k_1, ..., k_n are arbitrary integers. A *relation* between s_1, ..., s_n is a monomial r over S which is equal to id in the group G. Relations can also be written in the form $r_1 = r_2$, which is equivalent to $r_1 r_2^{-1} = \mathrm{id}$.

Definition 12. Let G be a transformation group, $S \subset G$ a subset and R a certain set of relations between the elements of S

We say that the set S is a *set of generators* and R a *set of defining relations* for the group G, if

1. any element of G can be represented as a monomial over S, and

2. any relation between the elements of S is a formal consequence of the relations belonging to the set R.

By a formal consequence of relations $r_1 = \mathrm{id}$, ..., $r_m = \mathrm{id}$, where every r_i is a monomial over the set S, we mean a new monomial $r = \mathrm{id}$ which can be deduced from the given set using the group operations (multiplication and taking the inverse) and their properties, such as associativity, the simplification rules $s^k s^l = s^{k+l}$, $s^0 = \mathrm{id}$, and the fact that the equalities $ab = \mathrm{id}$ and $ba = \mathrm{id}$ are equivalent. For example, the relations $ab^2 = \mathrm{id}$ and $ba^2 = \mathrm{id}$ imply that $b = a^{-2}$ and hence $a^3 = \mathrm{id}$.

Let us note that in a certain sense, the nature of the elements $s_1, \ldots, s_n$ is here irrelevant: later (page 136 in Chapter 5) we will give a definition of an abstract group with a prescribed set of generators and relations.

We pass to some exercises where the notion of defining relations is crucial.

Problem 32. *Suppose that A and B are two transformations that satisfy the relations*

$$A^3 = \mathrm{id}; \quad B^5 = \mathrm{id}; \quad AB = B^4A \tag{26}$$

and do not satisfy any relations that do not follow from these three by group axioms. Prove that the group generated by A and B is a cyclic group of order 3.

Solution. To simplify the formulas, both in the statement and the discussion of Problem 32 we omit the symbol of composition (small circle) and correspondingly use the word 'multiplication' instead of 'composition'.

An arbitrary element of the group G generated by A and B can be written as

$$B^{k_1}A^{l_1}B^{k_2}A^{l_2}\dots B^{k_m}A^{l_m},$$

where, in virtue of (26), we can assume that $0 \le k_i \le 4$ and $0 \le l_i \le 2$. Let us transform this 'word' using the following rule: *each time that an A appears next to a B on the left, replace AB by B^4A.* Applying this rule sufficiently many times, we will sooner or later push all B's to the left and arrive at a word of the type B^kA^l, where, as before, we are in a position to assume that $0 \le k_i \le 4$ and $0 \le l_i \le 2$.

Since the integer k may take 5 different values and the integer l three different values, we see that the total number of products B^kA^l is no greater than 15. It turns out, however, that not all of these elements are different. Indeed, we can deduce the following chain of equalities from the defining relations (26):

$$B = A^3B = A^2B^4A = AB^{16}A^2 = B^{64}A^3 = B^4,$$

where we have used the above described rule and the observation that *the letter A, when going through a B from left to right, multiplies its exponent by 4.* Therefore, $B^3 = \mathrm{id}$, which, together with the known identity $B^5 = \mathrm{id}$, implies that $B = \mathrm{id}$.

The group G is thus in fact generated by only one element A satisfying $A^3 = \mathrm{id}$. It remains to note that this generator A *cannot* be trivial, because the relation $A = \mathrm{id}$ is *not* a consequence of the relations (26). Indeed, if we take a 120 degree rotation for A and the identity transformation for B, then all the three relations (26) are true, while the relation $A = \mathrm{id}$ is false.

Exercise 66. Let A and B be two nontrivial movements of the plane such that $ABA^2 = \mathrm{id}$ and $B^2A = \mathrm{id}$. What is the order of the group generated by A and B? What kind of movements are A and B?

Similar arguments can be used in the following exercise, which at first glance looks totally unrelated to the theory of transformation groups.

Exercise 67. The language of the tribe Aiue has only 4 letters: A, I, U and E. The letter E is special. When used by itself, it means a certain word, but when added to any word in the beginning, the middle or the end, it does not change its meaning. Furthermore, each of the letters A, U, I, pronounced seven times in a row, makes a synonym of the word E. The following word fragments are also considered as synonyms: UUUI and IU, AAI and IA, UUUA and AU. The total number of people in the tribe is 400. Is it possible that all of them have different names?

Exercise 68. Find all pairs of generators in the group D_n. For each pair, indicate the defining relations.

The notion of defining relations can be used for groups that have any number of generators. For example, the cyclic group of order n is the group with one generator R and one defining relation $R^n = \mathrm{id}$. An infinite cyclic group is the group with one generator T which is not subject to any relations. We should like to emphasize once again the meaning of the last phrase. What we mean here is that there are no *non-trivial* relations for T, i.e., no relations that involve T and id and do not follow from the general properties of groups. Here is an example of a trivial relation: $T^{-1}T^4T^{-3} = \mathrm{id}$.

We shall now give an example of the group with three generators. Consider the group of plane movements which was denoted by G

in Problem 27 and which we will now denote by $p3m1$, its official crystallographic symbol (see Chapter 4 for a detailed discussion of crystallographic groups). By definition, the group $p3m1$ is generated by an infinite number of reflections, namely, the reflections in all the lines shown in Figure 1a.

It turns out, however, that this group is also generated by only three reflections S_a, S_b, S_c, where a, b and c are the sides of an equilateral triangle which forms the unit of the lattice shown in Figure 1a. To prove this fact, we have to show that the reflection S_x in any line x that belongs to the triangular lattice under study can be expressed in terms of S_a, S_b, S_c.

Consider, for example, the reflection S_l (notation of Figure 1a). Since the line l is symmetric to a with respect to c, we see that the movement S_l can be obtained from S_a using conjugation by S_c, i.e., $S_l = S_c \circ S_a \circ S_c$. Also, since $S_l(b) = m$, we have $S_m = S_l \circ S_b \circ S_l = S_c \circ S_a \circ S_c \circ S_b \circ S_c \circ S_a \circ S_c$.

A similar argument allows us to express any reflection S_x through S_a, S_b, S_c, because any line x belonging to the lattice can be obtained from the lines a, b, c by an appropriate series of reflections S_a, S_b, S_c.

We now observe that the three generators satisfy the following relations:

$$S_a^2 = S_b^2 = S_c^2 = \mathrm{id}, \tag{27}$$

$$(S_a \circ S_b)^3 = (S_b \circ S_c)^3 = (S_c \circ S_a)^3 = \mathrm{id}\,. \tag{28}$$

Exercise 69. Let F_1, F_2, F_3 be three plane movements that generate an infinite group and satisfy the relations

$$F_1^2 = F_2^2 = F_3^2 = (F_1 \circ F_2)^3 = (F_2 \circ F_3)^3 = (F_3 \circ F_1)^3 = \mathrm{id}\,.$$

Show that F_1, F_2, F_3 are reflections in the sides of an equilateral triangle.

This exercise shows that relations (27) and (28), supplied with the additional requirement of infiniteness, determine the group uniquely as the group of plane movements. Actually, it can be proved that these relations are the *defining relations* for the group $p3m1$ in the sense explained above.

Chapter 4

Arbitrary Groups

In this chapter we will introduce the general notion of a group, which includes transformation groups as a particular case. We will discuss the basic properties of groups in this general setting and consider some applications of groups in arithmetic.

1. The general notion of a group

In our study of transformation groups, it often did not matter that we dealt with *movements* or *transformations*. What mattered was the *properties* of the group G expressed as the following four requirements imposed on the *composition* defined for any pair of elements of G:

(1) the composition of two movements belonging to G also belongs to G;

(2) the composition of movements is associative;

(3) the group contains the identity transformation, which is characterized by the property that its composition with any movement, f is equal to f;

(4) together with every movement, the group contains the inverse movement.

We arrive at the general notion of a group, if we consider an arbitrary set, supplied with an operation which associates an element of this set with any pair of its elements and which enjoys a similar set of properties. For example, the set of all real numbers with the operation of addition $(\mathbb{R}, +)$ possesses all the listed properties and in this sense constitutes a group. We should like to draw the reader's attention to the fact that already in Chapter 1 (Problems 1, 7, etc.) we have used these properties for elements of various natures (numbers, points, vectors) and different operations (addition, multiplication). The analogy between Problem 32 and Exercise 67 is also noteworthy. All these facts testify that the notion of a group ought to be stated in a general setting. So here we go.

Definition 13. A *group* is a set G with the following properties:

(1) there is a rule (a *(binary) operation*), according to which for any ordered pair (a, b) of elements of the set G a certain element $a * b \in G$ is defined;

(2) the operation $*$ is associative, i.e. for any three elements $a, b, c \in G$ the following equality holds: $(a*b)*c = a*(b*c)$;

(3) in G, there is a *neutral* element, i.e., an element e such that $a * e = e * a = a$ for any $a \in G$;

(4) for every element $a \in G$ there is a *symmetric* element $a' \in G$ which satisfies $a * a' = a' * a = e$.

The four properties 1–4 are also called the *group axioms*. Note that any transformation group is a group in this general sense, although the definition of a transformation group (p. 76) consists of only two out of the four group axioms: number 1 and number 4. The reason is that axiom 2 always holds for the composition of transformations (p. 69), and axiom 3 follows from axioms 1 and 4, because the neutral element with respect to the composition of transformations is the identity transformation, and the question is only whether it belongs to the given set G.

Exercise 70. Prove that for a finite set G consisting of transformations axiom 4 follows from axioms 1–3.

Instead of the symbols $*$ (*asterisk*), $'$ (*prime*) and the terms 'neutral' and 'symmetric', whose meaning is explained in the definition of a group, other symbols and words are used in various specific circumstances:

- In case of transformation groups the group operation (composition) is denoted by a small circle ($\circ$), the neutral element is denoted by id (identity transformation), and the role of a symmetric element f' is played by the inverse transformation f^{-1}.
- For the group of numbers (integer, rational, real or complex) with the operation of addition the neutral element is 0 (zero), and the element symmetric to a given number a is the opposite number $-a$. (When we considered complex numbers as points in the plane, we called the neutral element the *pole* and denoted it by P.)
- For numeric groups with the operation of multiplication (for example, the set of all positive real numbers), the symbol of the operation is usually omitted, i.e. one writes ab instead of $a * b$, the neutral element is 1, and the symmetric element is the inverse number a^{-1}.

Groups that consist of numbers, vectors, etc., with the operation of addition, are referred to as *additive groups*. If the group operation is multiplication, then the group is called *multiplicative*.

The system of notation adopted for the multiplication of numbers is the most convenient, so it is often used for arbitrary groups. We shall also use it by default. The only thing that has to be kept in mind is that group multiplication, unlike numeric multiplication, is not in general commutative; hence ab and ba need not be the same element of the group.

Problem 33. *For each of the following sets with binary operations, determine whether it is a group:*

(1) *all even integers with the operation of addition;*

(2) *all odd integers with the operation of addition;*

(3) *all real numbers with the operation of subtraction;*

(4) *all natural numbers with the operation of addition;*

(5) *all non-negative integers with the operation of addition;*

(6) *all real numbers with the operation* $x * y = x + y - 1$.

Solution. In example 1 all the group axioms are obviously satisfied.

In example 2 the first axiom fails: the sum of two odd numbers is not an odd number.

In the third example the first axiom holds, but the second fails, because subtraction is not associative:

$$(6-5)-3 \neq 6-(5-3).$$

In the next example the first two axioms are fulfilled, but the operation does not have a neutral element: the only number that could play the role of the neutral element with respect to addition is zero, but it does not belong to the given set.

Example 5 differs from the previous one only in that the number 0 is added to the set, so that now the neutral element exists. But axiom 4 is not valid, because positive numbers do not have inverses in this set.

The last example requires more attention, because the operation is unusual. Let us check all group axioms one by one. It is clear that for an arbitrary pair (x, y) of real numbers $x + y - 1$ is also a real number, so that the first axiom holds. To verify the associativity, we have to calculate the two expressions $(x * y) * z$ and $x * (y * z)$, using the definition of $*$. We have

$$\begin{aligned} (x * y) * z &= (x + y - 1) * z = x + y - 1 + z - 1 = x + y + z - 2, \\ x * (y * z) &= x * (y + z - 1) = x + y + z - 1 - 1 = x + y + z - 2. \end{aligned}$$

We pass to the third axiom. The neutral element e must satisfy the identity $x + e - 1 = x$ for any x. This is true if and only if $e = 1$. Finally, since the operation $*$ is commutative, to determine the number symmetric to the given number x with respect to $*$, we have only one equation instead of two: $x * x' = 1$, i.e. $x + x' - 1 = 1$, whence $x' = 2 - x$. All the four axioms are thus satisfied, and the given set $(\mathbb{R}, *)$ is a group.

Exercise 71. Check whether the following sets with binary operations are groups. In case of a negative answer, indicate which of the four axioms fails.

(1) The set of all irrational numbers with the operation of addition.

(2) The set of all real numbers $x > 2$ with the operation $x * y = xy - x - y + 2$.

(3) The set of all binary rational numbers (i.e. fractions whose denominator is a power of 2) with the operation of addition.

(4) The set of all non-zero binary rational numbers with the operation of multiplication.

(5) Can you find a set of real numbers which forms a group with respect to the operation $x * y = (x + y)/(1 - xy)$?

We will indicate several simple but important corollaries of group axioms.

(1) The neutral element in a group is unique, i.e. there is only one element e that satisfies the requirements of the group axiom 3. Indeed, suppose that we have two elements e_1 and e_2 such that the following relation holds for every $a \in G$:

$$ae_1 = e_1a = ae_2 = e_2a = a.$$

Setting successively $a = e_1$ and $a = e_2$, we derive that $e_1 = e_1e_2 = e_2$.

(2) Any equation $ax = b$ is uniquely solvable in a group. This means that for any $a, b \in G$ there is a unique element $x \in G$ such that $ax = b$. Indeed, using group axioms 2 and 4, we can multiply the given equation by a^{-1} on the left and get $x = a^{-1}b$.

Exercise 72. Find a solution of the equation $xa = b$ and prove that it is unique.

(3) The previous assertions imply that:

- The inverse element for a given $a \in G$, defined by axiom 4, is unique.
- In every row and every column of the multiplication table of a group each element of the group appears exactly

once. One of these facts follows from the unique solvability of equations $ax = b$, the other from the unique solvability of equations $xa = b$.

(4) Group elements imply that the element $(a^{-1})^n$ is inverse to a^n. Therefore, as in the case of transformation groups, we can define zeroth and negative powers of the given element by setting $a^0 = e$, $a^{-n} = (a^{-1})^n$ for $n > 0$. Then, for arbitrary integer values of k and l we will have

$$a^k a^l = a^{k+l}. \tag{29}$$

(5) The last relation implies that the set of all integer powers of an element a forms a group. Such a group is called *cyclic*, and the element a is its *generator*. The *order* of the element a is defined as the smallest positive integer n such that $a^n = e$. If $a = e$, the order is 1 by definition; if a^n is different from e for every $n > 0$, we say that the order of a is infinite. In the latter case the subgroup generated by a is an *infinite cyclic group*. Note that the order (the number of elements) of the group generated by a is equal to the order of the element a.

(6) The axiom of associativity means that the product of three elements of a group, which involves two multiplications, does not depend on the order in which these multiplications are computed. Using induction, one can prove that this property is also true for any number of multiplications: any bracketing of the product $a_1a_2\dots a_n$ gives one and the same result. For example, $(a_1(a_2a_3))a_4 = ((a_1a_2)a_3)a_4 = (a_1a_2)(a_3a_4) = a_1(a_2(a_3a_4)) = a_1((a_2a_3)a_4)$.

So far, we have dealt with groups consisting of either transformations (transformation groups) or numbers (numeric groups). Now we will give an example of a group whose elements have quite a different nature.

Problem 34. *A* 3-switch *is an electric circuit with three inputs and three outputs connected by wires in such a way that every input corresponds to a certain output. The total number of 3-switches is six, and they are all displayed in Figure 1. The problem is to define a natural operation on the set of 3-switches which turns this set into a group.*

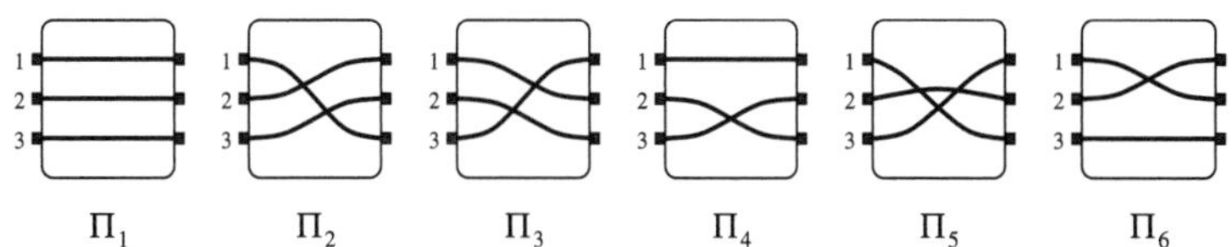

Figure 1. 3-switches

Solution. A natural operation on the set of switches is *concatenation.* To concatenate two switches means to connect the inputs of one of them to the outputs of another. For example, if we concatenate the switches Π_2 and Π_4, input number 1 of Π_2 will go through the input number 3 of Π_4 to the output number 2, so that in the result we have that input 1 is connected to output 2. Similarly, input 2 goes into output 1 and input 3 goes into output 3. This is the same pattern that we have for the switch Π_6. In this sense we can write that $\Pi_2\Pi_4 = \Pi_6$. Note that this operation is not commutative; for example, $\Pi_4\Pi_2 = \Pi_5$.

The complete multiplication (or, more exactly, concatenation) table for the set of 3-switches looks as follows:

	Π_1	Π_2	Π_3	Π_4	Π_5	Π_6
Π_1	Π_1	Π_2	Π_3	Π_4	Π_5	Π_6
Π_2	Π_2	Π_3	Π_1	Π_6	Π_4	Π_5
Π_3	Π_3	Π_1	Π_2	Π_5	Π_6	Π_4
Π_4	Π_4	Π_5	Π_6	Π_1	Π_2	Π_3
Π_5	Π_5	Π_6	Π_4	Π_3	Π_1	Π_2
Π_6	Π_6	Π_4	Π_5	Π_2	Π_3	Π_1

It turns out that the set of switches with this multiplication table forms a group. But how can one prove this? Using the direct procedure of verifying all the group axioms, just for axiom number 2 (associativity) one has to check $6^3 = 216$ equalities $\Pi_i(\Pi_j\Pi_k) = (\Pi_i\Pi_j)\Pi_k$. Fortunately, there is a less tedious way to do all these checks. Note that if you replace the letters Π_1, Π_2, Π_3, Π_4, Π_5, Π_6 in the multiplication table of the switches by id, R, R^2, S_a, S_c, S_b respectively and swap the two last rows and the two last columns of the table obtained, then you will get the multiplication table for the group D_3 (see p. 92). This means that the concatenation of 3-switches and the composition of the movements in the group D_3 establish exactly the same relations between the elements of the respective sets. Therefore, the operation of concatenation in the set of switches has all the properties that the composition of transformations has: it is associative, there is a neutral element (the switch Π_1), and every switch has an inverse. This implies that the set of 3-switches forms a group with respect to concatenation.

The mathematical content of the previous example consists in the description of all possible one-to-one mappings of the set $\{1, 2, 3\}$ into itself, or, in another terminology, all *transformations* of this set.

A transformation of the set $\{1, 2, \ldots, n\}$ is called a *permutation* of n elements, or a *permutation of degree* n. A permutation that takes 1 into i_1, 2 into i_2, ..., n into i_n is denoted as

$$\begin{pmatrix} 1 & 2 & \ldots & n \\ i_1 & i_2 & \ldots & i_n \end{pmatrix}.$$

There are $n!$ permutations of degree n in all, and they form a group, denoted by S_n.

There are even two different natural ways to define the group structure in the set of all permutations of a given degree. The first one is to treat permutations exactly like transformations and define the product $\sigma_1\sigma_2$ of the two permutations σ_1 and σ_2 as the composition $\sigma_1 \circ \sigma_2$ of the two mappings. We recall that the composition $\sigma_1 \circ \sigma_2$ is obtained by performing first the transformation s_2 and then the transformation s_1. According to this definition, we will have

$$\begin{pmatrix} 1 & 2 & 3 \\ 1 & 3 & 2 \end{pmatrix} \circ \begin{pmatrix} 1 & 2 & 3 \\ 3 & 2 & 1 \end{pmatrix} = \begin{pmatrix} 1 & 2 & 3 \\ 2 & 3 & 1 \end{pmatrix}.$$

Another way to define the product of the two permutations σ_1 and σ_2 is to first perform σ_1 and then σ_2 — exactly as we defined the concatenation of switches in Problem 34.

There are two schools of mathematicians: one maintains that the product of permutations should be defined as $\sigma_1 \circ \sigma_2$, the other, that it should be defined as $\sigma_2 \circ \sigma_1$. Multiplication tables for S_n adopted by the two schools differ by a reflection in the main diagonal. Actually, the two viewpoints are not so far apart: after studying the next section, you will be able to prove that the two permutation groups resulting from the two definitions are in fact *isomorphic*.

Here are some more problems where different groups appear, implicitly or explicitly.

Exercise 73. On a blackboard, several circles, squares and triangles are drawn. You are allowed to erase any two figures and replace them by a new figure following the rule:

— two circles make a circle;

— two squares make a triangle;

— two triangles make a square;

— a circle and a square make a square;

— a circle and a triangle make a triangle;

— a square and a triangle make a circle;

Prove that the shape of the last figure that will remain does not depend on the order in which the replacements are made.

Exercise 74. Consider the rational algebraic expressions in one variable, i.e., quotients of two polynomials in x with real coefficients. If A and B are two such expressions, then we can form the *superposition* $A * B$ by substituting B instead of x into A. Prove that the set Φ of all expressions that can be obtained by substitutions from $A_1 = 1 - x$ and $A_2 = 1/x$ forms a group, and find its order, its list of elements and its multiplication table.

Exercise 75. Find a rational expression B, different from a constant, such that $B * A_1 = B * A_2 = B$, where $A_1 = 1 - x$, $A_2 = 1/x$.

2. Isomorphism

When working on Problem 34, we deduced all the properties of the concatenation from the similar properties of the composition of transformations, using the fact that both operations have the same inner structure. The precise notion suited to characterize such situations is *isomorphism.*

Definition 14. Two groups G and H are said to be *isomorphic*, if there is a one-to-one correspondence, denoted by '$\leftrightarrow$', between the elements of G and the elements of H, such that $g_1 \leftrightarrow h_1$ and $g_2 \leftrightarrow h_2$ always imply $g_1g_2 \leftrightarrow h_1h_2$. One can also say that this correspondence *respects*, or *agrees with*, the group operations in both groups.

A more exact way to state the definition of an isomorphism is: the groups G and H are isomorphic, if there exists a one-to-one mapping $\varphi : G \to H$ such that

$$\varphi(g_1g_2) = \varphi(g_1)\varphi(g_2) \tag{30}$$

for any elements g_1 and g_2 of G. One can also say that φ is an *isomorphism* of the group G onto the group H.

At first sight it seems that the second version of the definition is different from the first, because the two groups do not enter symmetrically. In reality, however, the two groups have equal rights, because, if φ is an isomorphism of G onto H, then φ^{-1} will be an isomorphism of H onto G. Indeed, denoting $h_1 = \varphi(g_1)$, $h_2 = \varphi(g_2)$ and applying φ^{-1} to both sides of equation (30), we get $\varphi^{-1}(h_1)\varphi^{-1}(h_2) = \varphi^{-1}(h_1h_2)$.

The most direct way to establish the isomorphism, especially for finite groups, is to explicitly indicate all pairs of corresponding elements and then check that, when all elements of one group in its multiplication table are replaced by their counterparts from the other group, we obtain the multiplication table of the second group. (Of course, it might be necessary to change the order of columns and rows in the table obtained to make it literally coincide with the multiplication table of the second group as it was given.) Note that we followed this procedure in Problem 34.

Exercise 76. Check whether the following correspondence is an isomorphism between the group of 3-switches and the symmetry group of the equilateral triangle D_3: $\mathrm{id} \leftrightarrow \Pi_1$, $R^2 \leftrightarrow \Pi_2$, $R \leftrightarrow \Pi_3$, $S_b \leftrightarrow \Pi_4$, $S_c \leftrightarrow \Pi_5$, $S_a \leftrightarrow \Pi_6$.

This exercise leads to an important observation: if two groups G and H are isomorphic, the isomorphism $\varphi : G \to H$ is not in general unique. In particular, there might exist isomorphisms of a group onto itself, different from the identity transformation.

Exercise 77. Find all isomorphisms of the group D_3 onto itself.

The notion of an isomorphism illuminates the meaning of some analogies that an observant reader might have noticed in the material of the previous chapters. We can now state them as clear-cut problems.

Exercise 78. Prove that the set of all points of the plane with the operation of addition over a fixed pole (see p. 10) forms a group isomorphic to the additive group of plane vectors. Also prove that assigning to a point (or a vector) the pair of its coordinates in a certain basis establishes an isomorphism of the respective group and the group of pairs of real numbers with the operation defined by the rule

$$(a_1, b_1) + (a_2, b_2) = (a_1 + a_2, b_1 + b_2).$$

Exercise 79. Prove that the set of vertices of a regular hexagon with the multiplication described in Problem 7 (p. 26) forms a cyclic group isomorphic to the group C_6 of rotations with a common centre through angles which are multiples of 60°.

Exercise 80. Prove that the set {circle, triangle, square} with the operation defined in Exercise 73, is a group isomorphic to the cyclic group C_3. How many different isomorphisms between these groups are there?

Exercise 81. Prove that the group of rational algebraic expressions defined in Exercise 74 is isomorphic to the dihedral group D_3.

The results of Exercises 79 and 80 are generalized by the following assertion: *any two cyclic groups of the same order are isomorphic.* Indeed, let g be the generator of the first group G, and h the generator of the second group H. Define the mapping $\varphi : G \to H$ by the rule $\varphi(g^k) = h^k$. The law of multiplication of powers (29) holds in either group and implies that φ is an isomorphism:

$$\varphi(g^k g^l) = \varphi(g^{k+l}) = h^{k+l} = h^k h^l = \varphi(g^k)\varphi(g^l).$$

It is likewise clear that a group isomorphic to a cyclic group is itself cyclic, because the image of a generator under an isomorphism will also be a generator.

If we are interested in the inner structure of a group, we can forget about the nature of elements it consists of, keeping track only of the properties of the group operation.

Definition 15. An *abstract group* is a class of all groups which are isomorphic between themselves.

For example, all cyclic groups of order n, such as the group of rotations C_n or the group of complex n-th roots of unity, are representatives or, in Buddhist terminology, *incarnations*, of one and the same abstract cyclic group of order n. In the same way, the dihedral group D_3, the permutation group S_3, the group of switches (Problem 34) and the group of rational expressions (Exercise 74) are all representatives of one and the same abstract group. Later (page 136) we will explain how to define an abstract group with a given structure (set of relations between generators).

We now pass to the following general problem: *given two groups, decide whether they are isomorphic or not.* Usually, it takes more efforts to establish isomorphism than to establish non-isomorphism, because, in the first case, one normally has to *construct* the isomorphism, while in the second case, it is often enough to find some property which must be preserved by an isomorphism, but which distinguishes the groups under study. Here is a short list of some simple properties whose coincidence for two groups is a necessary condition for their isomorphism:

(1) *The order of the group.* Groups that have a different number of elements cannot be isomorphic.

Exercise 82. Is the group of all integers with addition isomorphic to the group of all even numbers with addition?

(2) *Commutativity.* A commutative group cannot be isomorphic to a non-commutative group.

(3) *Cyclicity.* A cyclic group cannot be isomorphic to a non-cyclic group.

Exercise 83. Are there any pairs of isomorphic groups in the list C_1, D_1, C_2, D_2, C_3, D_3, ...?

(4) *The orders of elements.* The number of elements of order n in one group must be equal to the number of elements of order n in the other group, because the orders of corresponding elements are the same.

We will only prove the last item, because it is somewhat more complicated than the others.

To begin with, note that under an isomorphism, the unit elements of the groups correspond to each other. In fact, if e is the unit of the group G and $\varphi : G \to H$ is an isomorphism, then $ee = e$ implies $\varphi(e)\varphi(e) = \varphi(e)$, which, upon multiplication by the element of H inverse to $\varphi(e)$, leads to the equality $\varphi(e) = e'$, where e' is the unit of H. Now let g be an element of G that has order n in G. By definition, $g^n = e$, whence $\varphi(g)^n = e'$, i.e., the order of $h = \varphi(g)$ does not exceed n. An inverse argument shows that the order of g cannot exceed the order of h. Therefore, the two orders are equal.

For example, to distinguish between the groups C_6 and D_3, it is enough to use any of the three criteria 2–4. First of all, C_6 is commutative, but D_3 is not. Also, C_6 is cyclic, but D_3 is not. Finally, C_6 has one element of order 1, one element of order 2, two elements of order 3 and two elements of order 6, while D_3 has one element of order 1, three elements of order 2 and two elements of order 3.

The list of properties that are necessary for two groups to be isomorphic is virtually infinite, because it contains any property of the group which can be formulated in terms of the group operation without referring to the specific nature of the elements of the group. We will, however, pass to the second half of the isomorphism problem. Suppose that we are given two groups G and H and we cannot find any intrinsic property that distinguishes them from each other. Then the conjecture arises that the groups are isomorphic. To prove this, one must construct an isomorphism $f : G \to H$ between G and H. How can this be done?

First we recall that, if e and e' are the unit elements in G and H respectively, then we have $\varphi(e) = e'$. Further, if $f(g) = h$ for a certain pair of elements $g \in G$ and $h \in H$, then by repeated application of (30) we can derive that $\varphi(g^k) = h^k$ for any natural k.

Exercise 84. Prove the equality $\varphi(g^k) = h^k$ for negative values of k.

We thus see that, if the mapping φ is defined on a certain element g of G, then it is also uniquely defined on the whole subgroup generated by g. Quite similarly, if the images of *several* elements $g_1, \ldots, g_n$ of the group G are known, then one can uniquely determine the image of any element expressible in terms of $g_1, \ldots, g_n$. If these elements are *generators* of G, then the values $\varphi(g_1) = h_1, \ldots, \varphi(g_n) = h_n$ completely determine the mapping φ. In the case of two generators we can write the corresponding formula as follows:

$$\varphi(g_1^{k_1} g_2^{l_1} \ldots g_1^{k_s} g_2^{l_s}) = h_1^{k_1} h_2^{l_1} \ldots h_1^{k_s} h_2^{l_s}. \tag{31}$$

Thus, if the group G is generated by two elements g_1 and g_2, then, to construct an isomorphism $\varphi : G \to H$, we must define its values $\varphi(g_1) = h_1$, $\varphi(g_2) = h_2$ and then extend the mapping to all of G according to relation (31). But how should the elements h_1, h_2 be chosen? It is clear that they ought to be a set of generators of the group H, they must have the same respective orders as g_1 and g_2, and they must satisfy all the relations that g_1 and g_2 satisfy. For example, if $g_1^2 g_2^3 = e$, then we must have $h_1^2 h_2^3 = e'$. These observations allow us to guess a 'candidate' for the isomorphism φ. After the mapping is constructed, one has to verify that it is really an isomorphism.

Problem 35. *Let ε be the complex number $-\frac{1}{2} + \frac{\sqrt{3}}{2} i$ (note that $\varepsilon^3 = 1$). Consider two functions of a complex variable $F_1(z) = \varepsilon z$,*

$F_2(z) = \overline{z}$. Prove that the set of all functions that can be obtained from F_1 and F_2 by superposition forms a group isomorphic to the dihedral group D_3.

Solution. We successively find that

$$\begin{aligned} F_3(z) &= F_1(F_2(z)) = F_1(\overline{z}) = \varepsilon\overline{z}, \\ F_4(z) &= F_2(F_2(z)) = F_2(\overline{z}) = z, \\ F_5(z) &= F_1(F_1(z)) = F_1(\varepsilon z) = \varepsilon^2 z, \\ F_6(z) &= F_2(F_1(z)) = F_2(\varepsilon z) = \overline{\varepsilon z} = \varepsilon^2\overline{z}. \end{aligned}$$

A straightforward check shows that further application of F_1 and F_2 to these expressions does not lead to any new functions. Thus, the set of six functions $F_1, \ldots, F_6$ is closed under superposition. The inverse of every function belonging to this list also belongs to this list. This proves that what we have is a group. The neutral element is the identity function F_4, the functions F_2, F_3, F_6 have order 2, and the functions F_1, F_5 have order 3. The group is not commutative since, for example, $F_1(F_2(z)) = F_3(z)$, while $F_2(F_1(z)) = F_6(z)$. These observations suggest that our group G is likely to be isomorphic to D_3.

To construct an isomorphism $\varphi : G \to D_3$, note that G by definition has two generators F_1 and F_2, whose orders are 3 and 2. In D_3 we can also find a system of two generators with orders 3 and 2: a rotation and a reflection. For example, set

$$F_1 \leftrightarrow R, \quad F_2 \leftrightarrow S_a$$

(in the notation of p. 92). Then

$$F_3 \leftrightarrow S_c, \quad F_4 \leftrightarrow \mathrm{id}, \quad F_5 \leftrightarrow R^2, \quad F_6 \leftrightarrow S_b.$$

Replacing every element of D_3 by the corresponding element of G in the multiplication table of D_3 (p. 92), we will obtain the following table:

	F_4	F_1	F_5	F_2	F_6	F_3
F_4	F_4	F_1	F_5	F_2	F_6	F_3
F_1	F_1	F_5	F_4	F_6	F_3	F_2
F_5	F_5	F_4	F_1	F_3	F_2	F_6
F_2	F_2	F_3	F_6	F_4	F_5	F_1
F_6	F_6	F_2	F_3	F_1	F_4	F_5
F_3	F_3	F_6	F_2	F_5	F_1	F_4

which, as one can easily check, is the correct multiplication table for G.

The isomorphism is thus established. There is, however, a more natural way to find an isomorphism between the two groups in question. Indeed, let us recall that a function of a complex variable can be viewed as an analytical representation of plane transformations. In particular, the function $F_1(z) = \varepsilon z$ corresponds to the rotations around 0 through 120°, while the function $F_2(z) = \overline{z}$, to a reflection in the real axis (axis a in Figure 2).

If we assign to every function obtained from F_1 and F_2 by superpositions the corresponding plane transformation, we will obtain a group isomorphism.[1]

The isomorphism constructed by the second method is called *natural.* A natural isomorphism reveals the reason why the two groups are isomorphic.

Exercise 85. Indicate a natural isomorphism

(1) between the group of 3-switches (see Problem 34) and the group D_3;

(2) between the group of rational algebraic expressions of Exercise 76 and the group D_3.

[1]Note that the terms 'superposition' and 'composition' actually have the same meaning, only the first one is used in analysis for functions, while the second one is used in geometry for transformations.

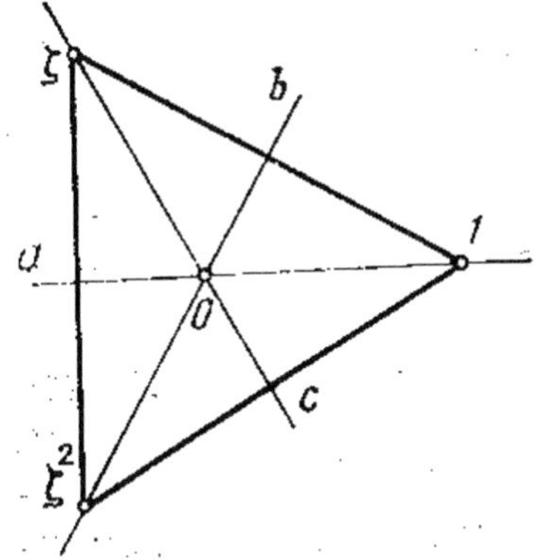

Figure 2. Equilateral triangle in the complex plane

We have to admit that the naturality we are talking about is by no means a strict mathematical notion, it rather bears a heuristic character. In a certain sense, any isomorphism is natural, but, in order to *understand* why is it natural, people often have to develop a special mathematical theory. A quest for natural isomorphisms that explain the reason why similar objects appear in different areas of mathematics is a powerful impetus that fosters the development of knowledge.

Here, we have used the general word 'knowledge' on purpose, to emphasize the fact that the notion of isomorphism is important not only in mathematics, but in any area of thinking. To get an idea of this general meaning of 'isomorphism', we invite the reader to contemplate over the following historical example.

In 1970, one of the problems of the entrance examination set for the Gelfand Correspondence Mathematical School in Moscow was published by two major journals in different formulations.

One of the journals stated the problem like this:

> One of three gangsters, known in city M under the names of Archie, Boss and Wesley, has stolen a bag of money. Each of them made three declarations:

- Archie:
  - I did not steal the bag.
  - On the day of the theft I was not in the city.
  - Wesley stole the bag.
- Boss:
  - Wesley stole the bag.
  - Even if I stole it, I would not confess.
  - I have lots of money.
- Wesley:
  - I did not steal the bag.
  - I've been long looking for a good bag.
  - Archie told the truth that he was not in the city.

During the investigation it was found that two declarations of each gangster were true and one false. Who stole the bag?

Another journal proposed the following problem (names mentioned belong to protagonists of Russian folk tales):

The King learned that somebody has killed the ruthless Dragon. He knew that this could only be done by one of the three famous warriors: Ilya Muromets, Dobrynia Nikitich or Alyosha Popovich. They were summoned to the King and each of them spoke three times. Here is what they said.

- I. M.:
  - I did not kill the dragon.
  - On that day I was travelling abroad.
  - A. P. did that.
- D. N.:
  - A. P. did that.
  - Even if I killed him, I would not confess.
  - There are many evil spirits still alive.
- A. P.:
  - I did not kill the dragon.

> – I've been long looking for a nice feat to do.
>
> – It is true that I. M. was abroad.
>
> The King found out that each of the three warriors twice told the truth and once lied. Who killed the dragon?

It is easy to see that, although the two problems are about quite different things, their logical structure is the same. Here is a glossary of names, things and actions that correspond to each other in the two problems:

Archie	Ilya Muromets
Boss	Dobrynia Nikitich
Wesley	Alyosha Popovich
bag	dragon
to steal	to kill
to leave the city	to go abroad

If, in the statement of the first problem, all the significant words are replaced by their counterparts from the right-hand column, the result almost coincides with the statement of the second problem — with the exception of Boss's third statement, which is actually irrelevant for the solution of the problem. In this sense, the two problems are isomorphic.

This isomorphism can be used as follows. If you solve the first problem and find that the answer is 'Boss', then you do not have to solve the second problem: the correct answer is given by the word that corresponds to Boss in our glossary, namely 'Dobrynia Nikitich'.

Figure 3. Isomorphism

In the same way one can use the isomorphism of groups: if G is isomorphic to H, then every assertion about G that can be stated in terms of the group operation will also hold for the group H, after an appropriate translation.

Another, more simple and direct, application of the group isomorphism is the computation of the product of elements in one group, using the product in another, provided that the second multiplication is less difficult and time-consuming. More exactly, if $\varphi : G \to H$ is an isomorphism between the groups $(G, *)$ and $(H, \circ)$, then the $*$-product can be computed by the formula

$$g_1 * g_2 = \varphi^{-1}(\varphi(g_1) \circ \varphi(g_2)). \tag{32}$$

A classical example of this kind of computations is provided by *logarithms*, invented by J. Napier[2] (early seventeenth century), who sought to replace the multiplication of numbers by a simpler operation — addition. Denoting the decimal logarithm of x by $\lg x$, we have the identity

$$x_1 x_2 = 10^{\lg x_1 + \lg x_2}, \tag{33}$$

which is a particular case of the general relation (32). The isomorphism that makes it possible to compute by using logarithms is the isomorphism $\lg : (\mathbb{R}_+, \cdot) \to (\mathbb{R}, +)$ of the group of positive real numbers with multiplication onto the group of all real numbers with addition. The two basic properties of the logarithmic function, namely

(1) it is on-to-one on the sets specified, and

(2) it satisfies the identity

$$\lg(x_1 x_2) = \lg x_1 + \lg x_2,$$

mean precisely that it establishes an isomorphism between the two groups.

Exercise 86.

(1) Find the analog of (33), if the decimal logarithm $y = \lg x$ is replaced by the Napier function $y = A \lg x + B$.

---

[2]Actually, logarithms in the contemporary sense of the word were introduced and tabulated by his disciple G. Briggs; Napier himself used a function $y = A \lg x + B$ with some constants A and B.

(2) Find a group operation $*$ on the set of all real numbers such that the Napier function gives an isomorphism of $(\mathbb{R}_+, \cdot)$ onto $(\mathbb{R}, *)$.

The second half of the last exercise is a particular case of the so-called *transition of structure.* Here is what we mean by that.

Let G be a group with operation $\triangle$ and H a set with no operation. Suppose that a one-to-one mapping $\varphi : H \to G$ is given. Then it is possible to *carry the group operation from G to H along φ* by the formula

$$h_1 \bigtriangledown h_2 = \varphi^{-1}(\varphi(h_1) \bigtriangleup \varphi(h_2)).$$

We have actually used this method:

- to derive the addition of points from the addition of vectors (Chapter 1);
- to define the unusual group operation over the real numbers $x * y = x + y - 1$ (Problem 33). This operation is obtained from the ordinary addition, which is carried over from one copy of $\mathbb{R}$ into another along the mapping $\varphi(x) = x - 1$. Indeed, $x * y = \varphi^{-1}(\varphi(x) + \varphi(y)) = ((x-1) + (y-1)) + 1 = x + y - 1$.

The operation $x * y = \dfrac{x+y}{1-xy}$ that appeared in Exercise 71 has the same origin. We have tried to perform the transition of the group structure $(\mathbb{R}, +)$ along the mapping $\varphi(x) = \tan x$, but we were not quite successful, because this mapping is not one-to-one. However, for any set $M \subset \mathbb{R}$ which is an additive group and has the property that the values of the tangent in the points of M are all different, the set of these values forms a group with respect to the operation $*$.

Exercise 87. Prove that:

(1) As such a set M one can take the set of all multiples of a real number α that is incommensurable with π.

(2) A set M with the required properties cannot contain any open interval of the real axis.

Exercise 88.

(1) What operation on real numbers is the result of transition of addition along the mapping $x \mapsto x^3$?

(2) How was the operation $x * y = xy - x - y + 2$ (Exercise 71) obtained?

In terms of the transition of structure, the notion of isomorphism can be formulated as follows: the mapping $\varphi : G \to H$ is an isomorphism of groups, if the group operation of H carried over to G along φ coincides with the group operation of G.

3. The Lagrange theorem

In this section we will state and prove the very first theorem of group theory, which was found by the French mathematician Lagrange in the late eighteenth century, even before the notion of group was explicitly introduced in mathematics in the nineteenth century by E. Galois.

Theorem 7 (**Lagrange**). *The order of a subgroup of a finite group is always a divisor of the order of the whole group.*

Since every element of a group generates a cyclic subgroup whose order is equal to the order of this element, we obtain, in particular, that *the order of a finite group is divisible by the order of each element.* The reader might have noticed this law in the examples considered above (groups C_{12}, D_3, etc.).

To prove the theorem of Lagrange in the general setting, we shall use the important construction of the *coset decomposition* of a group over a subgroup.

Let G be a group of order n, and H a subgroup of G of order m: $H = \{h_1, h_2, \ldots, h_m\}$. Since every subgroup contains the unit element of the group, we can assume that $h_1 = e$. Choose an arbitrary element g of G that does not belong to H, and consider the set

$$gH = \{gh_1, gh_2, \ldots, gh_m\}$$

obtained by multiplying all the elements of H by one and the same element g on the left. The set gH is called a *left coset of* G *over* H. It has two important properties:

(1) $|gH| = |H|$, i.e., gH has the same number of elements as H;

(2) $gH \cap H = \emptyset$, i.e., the sets gH and H do not have common elements.

To prove property (1), we have to show that all the elements of the list $gh_1, gh_2, \ldots, gh_m$ are distinct. Indeed, if we had $gh_i = gh_k$, then, after multiplying this equation by g^{-1} on the left, we would obtain $h_i = h_k$.

To prove (2), suppose that $h_i = gh_k$. This implies that g can be expressed as $h_i h_k^{-1}$ and must, therefore, belong to the subgroup H, contrary to the supposition.

The second property has the following generalization: *if $g_1 H$ is a coset and $g_2 \in G$ an element of the group that does not belong to $g_1 H$, then the two cosets $g_1 H$ and $g_2 H$ do not have common elements*, or, in other words, *two cosets either coincide or are disjoint.* In fact, if there were a common element, we would have $g_1 h_i = g_2 h_k$, hence $g_2 = g_1 h_i h_k^{-1}$ and, since $h_i h_k^{-1} \in H$, this would imply that $g_2 \in g_1 H$ and therefore $g_2 H = g_1 H$.

Now, the process of decomposing the group G into left cosets over the subgroup H can be described as follows. If the subgroup H coincides with the entire group G, then the coset decomposition consists of only one set, H. Otherwise, choose an element $g_1 \notin H$ and consider the coset $g_1 H$. If $H \cup g_1 H = G$, the process terminates. If not, we choose a $g_2 \in G$ which belongs neither to H nor to $g_1 H$ and thus obtain three pairwise disjoint cosets H, $g_1 H$ and $g_2 H$.

Since the group is finite, this process eventually terminates, and we obtain the required decomposition

$$G = H \cup g_1 H \cup \cdots \cup g_k H,$$

where each of the listed subsets has m elements and they are all pairwise disjoint.

Therefore, the number n of elements in the group is divisible by the number m of elements in the subgroup. The theorem is proved.

Let us remark that, similarly to the left coset decomposition, one can also consider the right coset decomposition. In general, these two decompositions do not coincide, and we will discuss this question in the next chapter.

Figures 4 and 5 show the left coset decompositions of the group D_3 over a subgroup of order 3 and a subgroup of order 2.

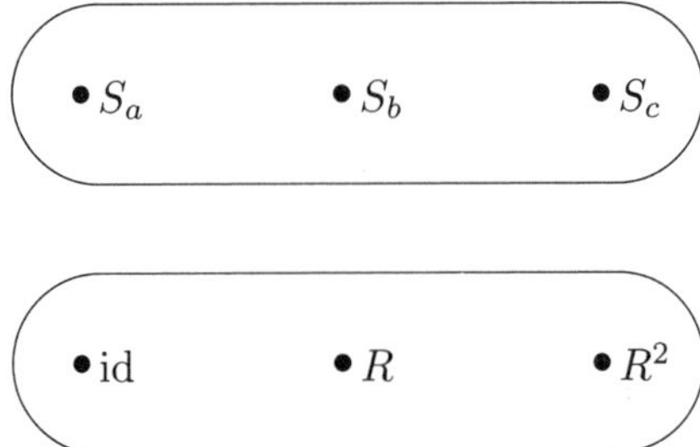

Figure 4. First coset decomposition of the group D_3

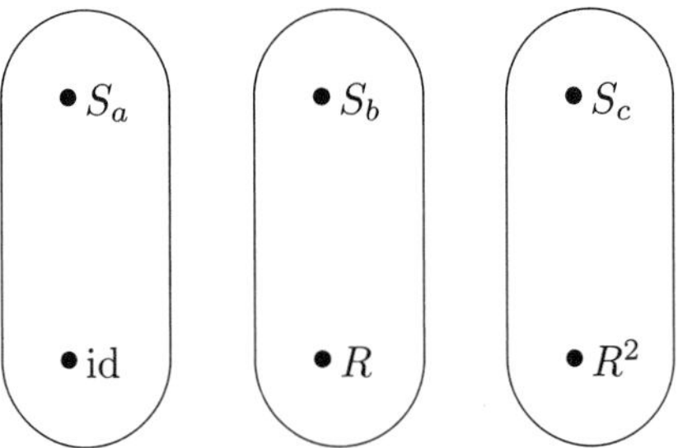

Figure 5. Second coset decomposition of the group D_3

Exercise 89. Find all subgroups of D_3.

The Lagrange theorem implies the following important fact.

Problem 36. *Prove that every finite group whose order is a prime number is cyclic.*

Solution. Let G be a group of prime order p and g an arbitrary element of G different from the unit element e. Denote by H the subgroup of G generated by g. The order of H is at least 2, since it contains e and g. The only divisor of the prime number p which is greater than 1 is p itself. Therefore, the order of H is p and $H = G$. The group G is thus generated by one element g.

As a direct consequence of the assertion just proved we obtain the following fact: *any group of prime order is commutative.*

We shall now discuss some applications of group theory, in particular, Lagrange's theorem, to arithmetic.

The simplest group that we come across in arithmetic is the group $\mathbb{Z}$ of all integers under addition. Since the group operation is addition, instead of powers of a certain element we will speak about its multiples, i.e., elements that are obtained by successively adding the given element to itself. The group $\mathbb{Z}$ is cyclic with generator 1, since every integer is a multiple of 1: $n = n \cdot 1$.

Exercise 90. Is there another generator in the group $\mathbb{Z}$?

As in any group, every element n of $\mathbb{Z}$ generates a subgroup. This subgroup consists of all multiples of n, and we denote it by $n\mathbb{Z}$.

Exercise 91. Prove that *every* subgroup of $\mathbb{Z}$ has the form $n\mathbb{Z}$ for a suitable n.

This simple result already provides a basis for useful applications in number theory. As an example, we will give a short proof of the following well-known fact: *if a and b are mutually prime numbers, then there exist two integers x and y such that $ax + by = 1$.*

Indeed, let H be the subgroup in $\mathbb{Z}$ generated by the two given numbers a and b. By definition, $H = \{ax + by \mid x, y \in Z\}$ (let us note that in multiplicative notation this expression would be written as $a^x b^y$). According to Exercise 91, we can find a natural n such that $H = n\mathbb{Z}$. Since the subgroup H contains the elements a and b, both of them are divisible by n. But, since they are mutually prime, n must be equal to 1. Therefore, the number 1 belongs to H and hence can be written as $ax + by$.

Lagrange's theorem does not directly apply to the pair consisting of the group $\mathbb{Z}$ and the subgroup $n\mathbb{Z}$, because these groups are infinite. However, the construction of the coset decomposition of the group over the subgroup does make sense and leads to the important notions of residue classes and modular arithmetic.

Take, for example, $n = 3$. If we add 1 to all the elements of the subgroup $3\mathbb{Z}$ (i.e., multiples of 3), we get the set of all whole numbers that have remainder 1 in division by 3. Likewise, adding 2 to all the elements of the subgroup, we get the set of all numbers that have remainder 2 in division by 3. Since there are no other remainders in division by 3, we see that the set of all integers splits into the 3 classes that we obtained. This is the *coset decomposition of* $\mathbb{Z}$ *over* $3\mathbb{Z}$. A visual representation of this decomposition is shown in Figure 6. Since the sets $3\mathbb{Z}$, $3\mathbb{Z}+1$ and $3\mathbb{Z}+2$ are infinite, only a few elements of each coset are listed.

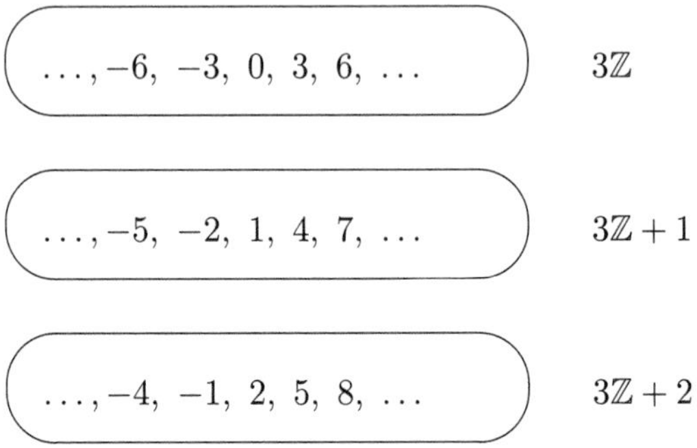

$\ldots, -6,\ -3,\ 0,\ 3,\ 6,\ \ldots$ $\quad 3\mathbb{Z}$

$\ldots, -5,\ -2,\ 1,\ 4,\ 7,\ \ldots$ $\quad 3\mathbb{Z}+1$

$\ldots, -4,\ -1,\ 2,\ 5,\ 8,\ \ldots$ $\quad 3\mathbb{Z}+2$

Figure 6. Coset decomposition of the group $\mathbb{Z}$ over $3\mathbb{Z}$

The cosets of $\mathbb{Z}$ over $m\mathbb{Z}$ are called *residue classes over* m. The class of all integers with remainder k in division by m, is conventionally denoted by $\overline{k}$. There are m residue classes in all; $\overline{0}$, $\overline{1}$,..., $\overline{m-1}$. For example, there are 3 classes modulo 3: $\overline{0}$, $\overline{1}$, $\overline{2}$ (see Figure 6).

Looking at the figure, one can see that the sum of two numbers always belongs to one and the same class, which is determined only by the classes of the two numbers under study and is independent of the particular choice of the representatives within the classes. For example, taking the representatives 1, -2, 7 of the class $\overline{1}$ and representatives -4, 5, 8 of the class $\overline{2}$, we see that the three sums $1+(-4)=-3$, $(-2)+5=3$, $7+8=15$ belong to one and the same class $\overline{0}$.

In general, the identity

$$(mx+k)+(my+l)=m(x+y)+(k+l)$$

allows us to define the operation of addition of residue classes over m: the sum of the classes $\overline{k}$ and $\overline{l}$ is the class that contains all the sums $k+l$, where k is a representative of the class $\overline{k}$ and l is a representative of the class $\bar{l}$. For example, if $m = 3$, we obtain the following addition table in the set of residues modulo 3:

$+$	$\bar{0}$	$\bar{1}$	$\bar{2}$
$\bar{0}$	$\bar{0}$	$\bar{1}$	$\bar{2}$
$\bar{1}$	$\bar{1}$	$\bar{2}$	$\bar{0}$
$\bar{2}$	$\bar{2}$	$\bar{0}$	$\bar{1}$

One can see from this table that the residue classes modulo 3 form a cyclic group of order 3.

Exercise 92. Prove that the residue classes over an arbitrary number m form a cyclic group of order m.

Can we also define the *multiplication* of residue classes in a similar way? Yes. Indeed, consider two classes $\bar{k}$, $\bar{l}$. The product of two arbitrary representatives $mx + k$ and $my + l$ of these two classes, equal to $m(mxy + xl + ky) + kl$, has the same remainder in division by m as the number kl. This remainder does not depend on the choice of the representatives. Therefore, the operation is correctly defined in the set of residue classes.

Here is the multiplication table of residue classes modulo 3:

$\times$	$\bar{0}$	$\bar{1}$	$\bar{2}$
$\bar{0}$	$\bar{0}$	$\bar{0}$	$\bar{0}$
$\bar{1}$	$\bar{0}$	$\bar{1}$	$\bar{2}$
$\bar{2}$	$\bar{0}$	$\bar{2}$	$\bar{1}$

It is evident that this multiplication does not satisfy all the group axioms, because it has one row and one column that entirely consist of zeroes, whereas in the multiplication table of a group no two elements may coincide. However, the smaller table that remains after the zero row and column are deleted,

$\times$	$\bar{1}$	$\bar{2}$
$\bar{1}$	$\bar{1}$	$\bar{2}$
$\bar{2}$	$\bar{2}$	$\bar{1}$

does obey all the group laws — it represents a cyclic group of order 2.

Exercise 93. Do all the nonzero residues modulo 6 form a multiplicative group?

This exercise suggests the following conclusion: *to form a multiplicative group out of residues over m it is reasonable to choose only numbers mutually prime with m.* For example, if $m = 6$, then the element $\overline{4}$, which has a common multiple 2 with the number 6, becomes $\overline{0}$ after multiplication by $\overline{3}$. But $\overline{0}$ cannot belong to a group, because it makes a whole row of zeroes in the multiplication table!

We shall now prove the following important fact: *the set of all residue classes $\overline{k}$ modulo m, such that the number k is mutually prime with m, forms a multiplicative group.*

Indeed, if two numbers are mutually prime with m, then their product is mutually prime, too; this means that the operation is closed on the given set. The associativity follows from the associativity of ordinary multiplication of numbers. The class $\overline{1}$ is mutually prime with m and plays the role of the unity. The only non-evident property that we must check is that every residue class in the set under study has an inverse. In other words, for every a, mutually prime with m, there must exist a number x, mutually prime with m and such that $ax \equiv 1 \pmod{m}$. The last formula is read aloud as '*the numbers ax and 1 are congruent modulo m*', which means, by definition, that ax has remainder 1 in division by m. This can be rephrased as follows: *there exists an integer y such that $ax + my = 1$.* Now recall that we have already proved this fact, stated in this form (as a corollary from Exercise 91). We did not mention that x will be mutually prime with m, but this is evident.

Exercise 94. Consider the set of two residue classes $\{\overline{2}, \overline{4}\}$ modulo 6. Does it make a group with respect to multiplication?

For any given m, we shall denote by $\mathbb{Z}_m^*$ the multiplicative group constituted by all residue classes modulo m which are mutually prime with m. The order of this group, i.e. the number of all such residue classes, is equal to $\varphi(m)$, the Euler function of m. We have encountered this function when we discussed the number of generators of a cyclic group (see page 89). We are now ready to apply the Lagrange theorem.

Note that in any group the following identity holds: $g^m = e$, where m is the order of the group, g an arbitrary element and e the unity. Indeed, if k is the order of g, then by Lagrange's theorem $m = kl$ for an appropriate integer l, and we have $g^m = (g^k)^l = e^l = e$. In the case of the group $\mathbb{Z}_m^*$ this implies the following theorem.

Theorem 8 (Euler). *If a is a number mutually prime with m, then*

$$a^{\varphi(m)} \equiv 1 \pmod{m},$$

where $\varphi(m)$ is the Euler function of m, i.e. the number of integers between 1 and m mutually prime with m.

In the case when $m = p$ is a prime number, we have $\varphi(p) = p - 1$, and the Euler theorem takes the shape of the following fact, known as *Fermat's little theorem.*

Theorem 9. *If p is a prime number, then*

$$a^{p-1} \equiv 1 \pmod{p} \tag{34}$$

for any integer a not divisible by p.

Historical remark. Neither Fermat nor Euler used group theoretic considerations explicitly to prove their theorems. It was only in the early nineteenth century that group theory came into being in the work of E. Galois. However, both Fermat and Euler *implicitly* did use such notions as, for example, coset decomposition of residue classes. These investigations became one of the sources from which group theory was born. The explicit application of group theory notions and theorems makes the arithmetical facts clearer and let us devise far-reaching generalizations.

To conclude this chapter, we propose two problems in elementary number theory that can be solved using residues and Euler's theorem.

Exercise 95. Prove that the equation $x^2 = 3y^2 + 8$ has no integer solutions.

Exercise 96. What are the two last digits of the number 2003^{2004}?

Chapter 5

Orbits and Ornaments

Topics that will be touched upon in this chapter include group actions, orbits, invariants, and ornaments.

Transformation groups, by their definition, *act* on certain sets. Thus, the group of movements of the plane acts on (the set of all points of) the plane. The permutation group S_3 acts on the set $\{1,2,3\}$. The ability to transform the sets is so inherent in the notion of a group that it is also preserved for arbitrary abstract groups. To give the exact definition of the group action, we shall need the notion of a homomorphism.

1. Homomorphism

The notion of a homomorphism generalizes that of an isomorphism. A homomorphism is defined by the same operation-preserving property, only without the requirement that it must be a one-to-one correspondence.

Definition 16. A *homomorphism* from a group G into a group H is a mapping $\varphi : G \to H$ such that

$$\varphi(ab) = \varphi(a)\varphi(b) \tag{35}$$

for all $a, b \in G$ (the group operation is here written as multiplication, but of course, it may have an arbitrary nature).

An isomorphism is thus a one-to-one homomorphism. Some properties of isomorphisms generalize to arbitrary homomorphisms. Thus, the image of the unit $e \in G$ under a homomorphism is always the unit $e' \in H$. Also, for any element $g \in G$ we always have

$$\varphi(g^{-1}) = \varphi(g)^{-1}. \tag{36}$$

Both of these equalities easily follow from (35).

Note that for any two groups G and H the mapping that takes all the elements of G into the unit of H is a homomorphism, called the *trivial homomorphism.*

If $\varphi : G \to G'$ is an isomorphism and G' is a subgroup of a bigger group H, then we can view φ as a mapping from G into H which is, of course, a homomorphism. Such homomorphisms are referred to as *injective*, or *monomorphic homomorphisms.* They are characterized by the property that no two different elements of G go into the same element of H under φ.

The image of G by the trivial homomorphism is the trivial subgroup (consisting of only one element). Note that the image of G by any homomorphism $\varphi : G \to H$ is a subgroup of H. The most interesting class of homomorphisms, in a certain sense dual to the class of monomorphisms, consists of *epimorphic*, or *surjective homomorphisms.* A homomorphism $\varphi : G \to H$ is *surjective*, if its image coincides with the whole of H, or, which is the same, if every element of H is the image of some element of G.

We will consider several examples of surjective homomorphisms.

Problem 37. *Construct a surjective homomorphism of the group of integers $\mathbb{Z}$ with the operation of addition onto the group of residue classes $\mathbb{Z}_m$.*

Solution. The solution is very simple: the mapping p that takes every number a into $\overline{a}$, the class of a modulo m, is the required epimorphism. It preserves the addition, because by definition of the group $\mathbb{Z}_m$ (see p. 122) we have

$$p(a+b) = \overline{a+b} = \overline{a} + \overline{b} = p(a) + p(b).$$

Also, the mapping p is obviously surjective.

Note in passing that p also preserves the product: $p(ab) = p(a)p(b)$, so it is a homomorphism with respect to multiplication — and you probably used both of these properties while solving Exercise 95.

Exercise 97. For what values of m and n does there exist a homomorphism of $\mathbb{Z}_m$ onto $\mathbb{Z}_n$?

Problem 38. *Find a homomorphism of S_4 onto S_3.*

Solution. Recall (1) that S_n denotes the permutation group of n elements. Thus, S_4 consists of 24 permutations of the set of cardinality 4, while S_3 is made up of 6 permutations of the set of cardinality 3. We have seen earlier (Exercise 74) that S_3 is isomorphic to the group of functions Φ generated by $1/x$ and $1 - x$. Therefore, it is enough to construct a homomorphism of S_4 onto Φ. This can be done as follows.

Consider the expression in four variables

$$x = \frac{a-c}{b-c} : \frac{a-d}{b-d},$$

which is called the *double ratio* of a, b, c and d. If the four letters a, b, c, d in this expression are permuted, the value of the double ratio is changed. It is remarkable that the new value can always be expressed in terms of x alone. For example, after the permutation $\begin{pmatrix} a & b & c & d \\ b & c & d & a \end{pmatrix}$ (i.e., when $a \mapsto b$, $b \mapsto c$, $c \mapsto d$, $d \mapsto a$) we obtain

$$\frac{b-d}{c-d} : \frac{b-a}{c-a} = \frac{x}{x-1}.$$

Exercise 98. For each of the 24 permutations of the letters a, b, c, d, find the expression of the double ratio after the permutation in terms of its initial value x.

A diligent reader who has solved this exercise will recall that the set of 6 functions obtained is exactly the group of functions that we know as Φ. Denote the rational function that corresponds to the permutation σ

by $f_\sigma(x)$. To show that the mapping $f : \sigma \mapsto f_\sigma(x)$ is a homomorphism, we have to check that $f_{\tau\sigma} = f_\tau \circ f_\sigma$. Indeed, if

$$f_\sigma(x) = \frac{\sigma(a) - \sigma(c)}{\sigma(b) - \sigma(c)} : \frac{\sigma(a) - \sigma(d)}{\sigma(b) - \sigma(d)} = y,$$

then evidently

$$f_{\tau\sigma}(x) = \frac{\tau\sigma(a) - \tau\sigma(c)}{\tau\sigma(b) - \tau\sigma(c)} : \frac{\tau\sigma(a) - \tau\sigma(d)}{\tau\sigma(b) - \tau\sigma(d)} = f_\tau(y),$$

and thus $f_{\tau\sigma}(x) = f_\tau(f_\sigma(x))$.

Now we would like to draw your attention to the following interesting observation. The fact that f is a homomorphism, i.e. satisfies the relation $f_{\tau\sigma} = f\tau \circ f\sigma$, greatly simplifies the solution of Problem 38. In fact, to prove that the image of f coincides with the set Φ, we do not have to check all the 24 permutations of the four letters as suggested in the exercise above. It is sufficient to check 3 transpositions, $a \leftrightarrow b$, $b \leftrightarrow c$ and $c \leftrightarrow d$, which generate the group S_4. These transpositions correspond to the functions $1/x$, $1-x$ and $1/x$, respectively, and we know that the two functions $1/x$ and $1-x$ generate the group Φ.

Exercise 99. Assign $+1$ to every movement that preserves orientation and -1 to every movement that changes orientation. Check that this assignment is a homomorphism of the group of all plane movements onto the group of 2 elements $\{+1, -1\}$ with the operation of multiplication.

Problem 39. *Construct a homomorphism of the group G of all proper plane movements onto the group T of complex numbers whose modulus is 1.*

Solution. Recall (8) that any proper plane movement can be written analytically as a complex function $f(z) = pz + a$, where $|p| = 1$. We define the mapping $\varphi : G \to T$ by $\varphi(f) = p$. Let us check that this mapping is a group homomorphism. Indeed, the composition of two movements, f and g defined by $g(z) = qz + b$, is

$$g(f(z)) = q(pz + a) + b = qpz + (aq + b),$$

where the coefficient of z is qp. Therefore, $\varphi(gf) = \varphi(g)\varphi(f)$.

Geometrically, the assertion just proved implies that when two rotations, even with different centres, are multiplied, the angles of rotation are added together. In particular, the movement $R_A^{\varphi} \circ R_B^{-\varphi}$ is always a parallel translation.

To solve the next exercise, you will find the following fact useful: *the angle between the lines l and m is equal to the angle between $R_A^{\varphi}(l)$ and $R_B^{\varphi}(m)$.* Indeed, rotations preserve the angles; hence the angle between l and m equals the angle between $R_A^{\varphi}(l)$ and $R_A^{\varphi}(m)$ — and the two lines $R_A^{\varphi}(m)$ and $R_B^{\varphi}(m)$ are parallel.

Exercise 100. Let BE and CF be the altitudes of the triangle ABC, and O the centre of the circumscribed circle. Prove that $AO \perp EF$.

Exercise 101. The *determinant* of a matrix $\begin{pmatrix} a & b \\ c & d \end{pmatrix}$ is the number $ad-bc$. The *product* of two matrices $\begin{pmatrix} a & b \\ c & d \end{pmatrix}$ and $\begin{pmatrix} m & n \\ p & q \end{pmatrix}$ is the matrix $\begin{pmatrix} am+bp & an+bq \\ cm+dp & cn+dq \end{pmatrix}$. Prove that the set of all matrices with non-zero determinant forms a group, and the determinant is a homomorphism of this group onto the group of non-zero numbers with multiplication.

2. Quotient group

Let us look at the group of residue classes $\mathbb{Z}_m$ and the homomorphism $\mathbb{Z} \to \mathbb{Z}_m$ from a more general viewpoint. The group $\mathbb{Z}_m$ consists of the cosets of the group $\mathbb{Z}$ modulo the subgroup $m\mathbb{Z}$. This fact can also be stated as *$\mathbb{Z}_m$ is the quotient group of $\mathbb{Z}$ modulo $m\mathbb{Z}$*, which is written as $\mathbb{Z}_m = \mathbb{Z}/m\mathbb{Z}$.

Now let G be an arbitrary group and H a subgroup of G. Consider the set of cosets in G modulo H and try to make it into a group, using the construction of $\mathbb{Z}_m$ as an example. The first question that arises is: what cosets should we use, left cosets eH, g_1H, g_2H, ... or right cosets He, Hg_1, Hg_2, ...? This question does not arise for the group $\mathbb{Z}$, since it is commutative.

Recall that the left coset gH is made up of all products gh where $g \in G$ is fixed and h ranges over H, whereas the right coset Hg consists of all products hg. If the group is not commutative, then in general $gH \neq Hg$. For example, take the group of plane movements D_3 as the whole group G, and the subset $H = \{\mathrm{id}, S_a\}$ as the subgroup. Then, using the multiplication table (p. 92), we find that

$$\begin{aligned} RH &= \{R \circ \mathrm{id}, R \circ S_a\} = \{R, S_b\}, \\ HR &= \{\mathrm{id} \circ R, S_a \circ R\} = \{R, S_c\}. \end{aligned}$$

However, for the subgroup $C_3 = \{\mathrm{id}, R, R^2\}$ both the left and the right coset decompositions consist of the same two classes: the subgroup C_3 itself and its complement $D_3 \setminus C_3$.

Definition 17. A subgroup H of a group G is called *normal*, if for any element $g \in G$ the two cosets gH and Hg coincide.

If $H \subset G$ is normal, then the following train of equalities is true:

$$(g_1H)(g_2H) = g_1(Hg_2)H = g_1(g_2H)H = g_1g_2H,$$

meaning that if you choose any element of the coset g_1H and any element of the coset g_2H, then their product will belong to one and the same coset $(g_1g_2)H$. Therefore, in the set of cosets over a normal subgroup there is a well-defined multiplication: $\bar{g}_1\bar{g}_2 = \overline{g_1g_2}$, where the bar over a letter denotes the coset of an element over the given normal subgroup: $\bar{g} = gH = Hg$. It is evident that this operation is associative; the role of unit element is played by the coset $H = eH$, and the element inverse to gH is $g^{-1}H$. The set of cosets H, g_1H, g_2H, ... thus acquires the structure of a group.

Definition 18. Let G be a group with a normal subgroup H. The *quotient group of* G *over* H, denoted by G/H, is the set of all cosets over H with the product of two cosets given by the rule $\bar{g}_1\bar{g}_2 = \overline{g_1g_2}$.

The structure of the quotient group G/H can be read off the multiplication table of G, if the elements in the first row and and the first column of the table are arranged by cosets. Look, for example, at the multiplication table of the group D_3 (p. 92). You can clearly see that the table is split into 4 blocks of the size 3×3 each. Denoting the block that corresponds to rotations by R and the block that corresponds to reflections by S, we can represent the block structure of the whole table as

	R	S
R	R	R
S	S	R

This table defines a cyclic group of order 2 isomorphic to $\mathbb{Z}_2$. Thus, we can write $D_3/C_3 \cong \mathbb{Z}_2$, where the sign $\cong$ stands for isomorphism.

Problem 40. *Let G be the group of proper movements of the plane, consisting of all rotations and all translations. Denote by N the subgroup of rotations around a fixed point A and by K the subgroup of all translations. Determine whether these subgroups are normal and, if appropriate, find the structure of the quotient groups.*

Solution. The condition of normality $gH = Hg$ can be rewritten as $gHg^{-1} = H$, which means that a subgroup H is normal if and only if, together with every element of it, it contains all the conjugate elements. Since conjugation is looking at things from a different viewpoint (see p. 82), a subgroup is normal whenever it looks the same from any standpoint. It is clear that, for a person living in the plane, the set of all translations looks the same, whatever his position may be. However, the set of rotations with a fixed centre A looks different for a person placed at A and for a person placed somewhere else. Therefore, the subgroup K is normal, but N is not.

A more rigorous proof of this fact can be obtained using the result of Exercise 58 (p. 234). The movement conjugate to a translation is always a translation; therefore K is normal. The movement conjugate to a rotation around A by a movement f is a rotation around $f(A)$; therefore N is not normal.

Since $K \subset G$ is normal, the quotient group G/K is defined. To understand its structure, consider the homomorphism $\varphi : G \to T$ studied above in Problem 39. We claim that two elements of G have the same image under φ if and only if they belong to the same coset of G with respect to K. Indeed, let $\varphi(f) = \varphi(g) = p \in T$. If $p = 1$, then both f and g are translations and belong to K. If $p = \cos\alpha + i\sin\alpha \neq 1$, then both f and g are rotations through the angle α, say, $f = R_A^\alpha$ and $g = R_B^\alpha$. In this case $f = R_A^\alpha = R_B^\alpha \circ (R_B^{-\alpha} \circ R_A^\alpha) \in gH$, because $R_B^{-\alpha} \circ R_A^\alpha$ is a translation. Conversely, if f and g belong to one coset over K, then $f = g \circ h$ and $\varphi(f) = \varphi(g)$.

The property that we have just proved implies that φ establishes a one-to-one correspondence between the sets G/K and T, which we can denote by $\bar{\varphi}$. The mapping φ obviously agrees with the group operations and thus yields an isomorphism $\varphi : G/K \to T$. The quotient group G/K is thus the same thing as the group of complex numbers with unit modulus, or the group of all rotations with a common centre.

Note that the coset decomposition of G over K has a simple geometric meaning: every coset consists of rotations through the same angle around an arbitrary centre. The subgroup K itself (the set of all translations) is the unit coset — it corresponds to the unit of the group T under φ.

Generalizing this argument, we arrive at the following assertion, called the *first homomorphism theorem.*

Theorem 10. *If φ is a homomorphism of a group G onto a group H and $K \subset G$ is its kernel, i.e., the set of all elements of G that go into the unit of H under φ, then $G/K \cong H$.*

Note that the kernel K of any homomorphism φ is always a normal subgroup in G, so that the quotient G/K is correctly defined. Indeed, if $k \in K$, then $\varphi(k) = e$; hence $\varphi(gkg^{-1}) = \varphi(g)\varphi(k)\varphi(g^{-1}) = \varphi(g)e\varphi(g^{-1}) = e$, which means that $gkg^{-1} \in K$.

The first homomorphism theorem implies that the orders of the three groups G, H and K are related by the simple equality $|G| = |H| \cdot |K|$. In particular, we have the following corollary: *if there exists* *a homomorphism of finite groups $G \to H$, then the order of H is a divisor of the order of G.*

Exercise 102. Is there a homomorphism (a) of D_3 onto $\mathbb{Z}_2$? (b) of D_3 onto $\mathbb{Z}_3$?

Problem 41. *Find the kernel of the homomorphism $\varphi : S_4 \to \Phi$ discussed in Problem 38.*

Solution. Let $K \subset S_4$ be the kernel of φ. Since φ is a surjective homomorphism (i.e., its image coincides with the entire group Φ), the groups S_4/K and Φ are isomorphic and hence $|K| = 24 : 6 = 4$. It is easy to check that the four permutations

$$\begin{pmatrix} a & b & c & d \\ a & b & c & d \end{pmatrix}, \quad \begin{pmatrix} a & b & c & d \\ b & a & d & c \end{pmatrix},$$
$$\begin{pmatrix} a & b & c & d \\ c & d & a & b \end{pmatrix}, \quad \begin{pmatrix} a & b & c & d \\ d & c & b & a \end{pmatrix}$$

leave invariant the expression

$$x = \frac{a-c}{b-c} : \frac{a-d}{b-d}.$$

The multiplication table for these four elements coincides, up to the choice of notation, with the multiplication table of the group D_2. Thus $K \cong D_2$, and the homomorphism theorem in our example can be written as $S_4/D_2 \cong D_3$.

Exercise 103. Find the image and the kernel of the homomorphism of the additive group of functions $f : \mathbb{R} \to \mathbb{R}$ into itself given by the formula $f(x) \mapsto f(x) + f(-x)$.

Exercise 104. Let S be the group of rotations of the plane with a fixed centre, and C_n its cyclic subgroup of order n. Prove that $S/C_n \cong S$.

Exercise 105. Using the first homomorphism theorem, represent the group S of the previous exercise as a quotient group of $(\mathbb{R}, +)$.

3. Groups presented by generators and relations

We are now in a position to explain the construction of an abstract group with a given set of generators and defining relations. Before (page 93 in Chapter 3) we have already talked about generators and relations in a given, already defined group. The current problem is the inverse: now we want to *define* a group, starting from an arbitrary set of generators and relations between them.

This definition relies on the notion of the quotient group that we have just studied and the notion of a free group that we will now define.

Definition 19. Let S be an arbitrary set, consisting of letters in a certain alphabet. The *free group over* S, denoted by $F(S)$, consists of all monomials over S (see page 93), the group operation being given by writing two monomials side by side and using the simplification rules $s^k s^l = s^{k+l}$ and $s^0 = 1$.

For example, if S consists of one element, then $F(S)$ is the infinite cyclic group.

Definition 20. Let R be a set of monomials over S (each monomial r is understood as a relation $r = 1$ between the elements of S). The *group with generators and defining relations* R is defined as the quotient group $F(S)/H(R)$, where $F(S)$ is the free group over S and $H(R)$ is the minimal normal subgroup of $F(S)$ containing all relations belonging to R (in other words, $H(R)$ is the subgroup of $F(S)$ generated by all elements conjugate to the elements of R).

An abstract group with generators $s_1, \dots, s_n$ and relations $r_1, \dots,$ r_m is denoted by

$$\langle\, s_1, \dots, s_n \,|\, r_1, \dots, r_m \,\rangle$$

(The relations on the right are sometimes written simply as monomials over S, each monomial r meaning the equality $r = 1$).

For example, it is easy to see that

$$\langle\, a \,|\, a^n = 1 \,\rangle$$

is the cyclic group of order n.

Problem 42. *Prove that the group*

$$\langle\, a, b \,|\, ab = 1 \,\rangle$$

is the infinite cyclic group, isomorphic to the additive group of all integers $\mathbb{Z}$.

Solution. The free group $F(a,b)$ consists of all words $a^{k_1}b^{l_1}\ldots a^{k_n}b^{l_n}$ of arbitrary length and with arbitrary integer exponents. To obtain the quotient group, such words should be considered modulo the elements of the subgroup $H(R)$: if $x = yh$ or $x = hy$, where $h \in H(R)$, then x and y belong to one and the same coset. Now, $a = (ab)b^{-1}$; therefore, a is equivalent to b^{-1}, and hence every word in a and b is equivalent to some word in b only. We see that every element of the quotient group $F(S)/H(R)$ is a power of $\bar{b}$, so that this group is cyclic.

Note that in every element of the subgroup $H(R)$ the sum of all exponents of a is equal to the sum of all exponents of b; therefore b^n with $n \neq 0$ cannot belong to $H(R)$, which means that all powers of $\bar{b}$ are different. The group $F(S)/H(R)$ is thus infinite.

Remark. In the above argument, the generators a and b play symmetrical roles, because the product ba is conjugate to ab: $ba = a^{-1}(ab)a$ and therefore belongs to the subgroup $H(R)$ — this is one of the reasons why in Definition 20 the subgroup $H(R)$ is set to be the *normal* subgroup generated by R.

Exercise 106. What is the group presented by

$$\langle\, a, b \,|\, a^2 = 1,\ b^n = 1,\ aba = b^{-1} \,\rangle\,?$$

Exercise 107. Prove that

$$\langle\, a, b \,|\, aba = bab \,\rangle \cong \langle\, x, y \,|\, x^2 = y^3 \,\rangle\,.$$

4. Group actions and orbits

Group actions can be defined in terms of homomorphisms as follows.

Definition 21. Let G be a group and X a set. An *action* of G on X is a homomorphism of G into the group of transformations of X (see p. 76):

$$T : G \to \mathrm{Tr}(X).$$

The image of a point $x \in X$ under the action of the transformation T_g is $T_g(x)$, which is often denoted by gx for short.

We should like to stress from the beginning that a given group G can act on a given set X in a variety of different ways, because in general there may exist many different homomorphisms $G \to \mathrm{Tr}(X)$. For example, the symmetry group of an equilateral triangle D_3 acts on the plane by definition. However, the action depends on the choice of the equilateral triangle in question, or more exactly, on the choice of its centre and three symmetry axes.

Figure 1 shows an equilateral triangle with centre O and symmetry axes a, b, c.

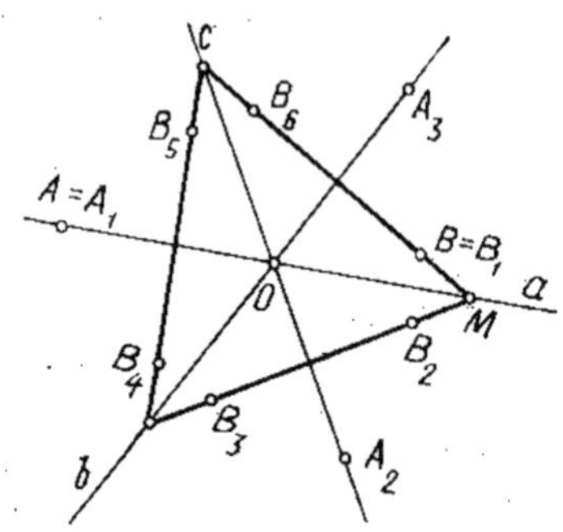

Figure 1. Action of the group D_3 on the plane

Under the action of any element of D_3, the origin (point O) goes into itself. A point lying on one of the lines a, b, c and different from O gives three different points (including itself). Any other point of the plane gives six different points under the action of the group.

Definition 22. The set of all points obtained from a given point x under the action of a group is called the *orbit* of this point:

$$\mathcal{O}(x) = \{T_g(x) \,|\, g \in G\}.$$

The cardinality of the set $\mathcal{O}(x)$ is called the *length* of the orbit.

In the previous example we had orbits of three different kinds: of lengths 1, 3 and 6.

It is clear that any point always belongs to its own orbit, because it goes into itself under the action of the neutral element of the group. A point whose orbit consists of only one point is called a *fixed point* of the action. In our example the only fixed point was the point O.

Look at Figure 1. The orbit of the point $A = A_1$ consists of three points A_1, A_2, A_3. If you take A_2 or A_3 as the initial point, you will get the same orbit. This is a manifestation of the general property: *the orbit of any point belonging to the orbit of a point x coincides with* $\mathcal{O}(x)$. Indeed, if $y \in \mathcal{O}(x)$, then $y = hx$ for a certain group element $h \in G$. Then $\mathcal{O}(y) = \{gy|g \in G\} = \{ghx|g \in G\}$, but the axioms of the group imply that, for any fixed element $h \in G$, the set of all products gh, where g ranges through G, coincides with the set G.

This observation implies the following important fact: *any two orbits either coincide or do not have common elements.* In fact, if two orbits $\mathcal{O}(x)$ and $\mathcal{O}(y)$ have a common element z, then $\mathcal{O}(x) = \mathcal{O}(z) = \mathcal{O}(y)$.

The meaning of the word *orbit* and the splitting of the set into orbits can be very clearly seen for the action of the circle S, understood as the multiplicative group of complex numbers with modulus 1, on the complex plane by means of multiplication (Figure 2). This action has one fixed point (the number 0), while the rest of the plane splits into concentric circles which are the orbits.

Exercise 108. Let D_3 be the symmetry group of an equilateral triangle in the complex plane, whose centre is at the origin and one of whose symmetry axes is the x-axis. Consider the action of this group on the finite set $\{0, 1, -1, 2, -2, i\sqrt{3}, -i\sqrt{3}, 4, -4, 2 + 2i\sqrt{3}, 2 - 2i\sqrt{3}, -2 + 2i\sqrt{3}, -2 - 2i\sqrt{3}\}$. Find the orbits of this action.

Problem 43. *Define a natural action of the group of rational expressions Φ (Exercise 74) on a suitable set.*

Solution. A rational expression in one variable can be considered as a function, i.e. as a mapping of the real line

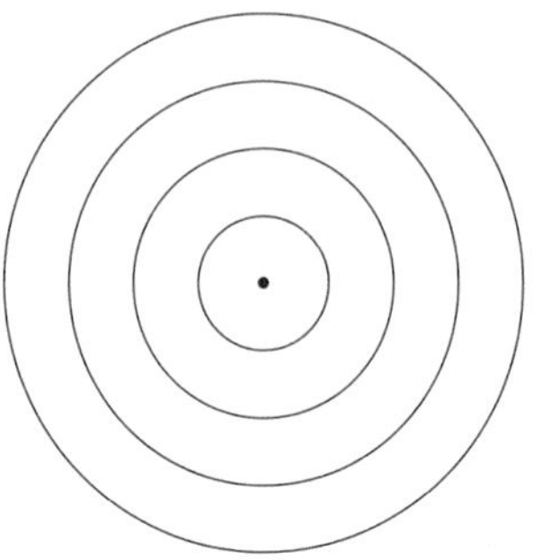

Figure 2. Orbits of the group S

into itself. Unfortunately, these mappings are not defined at all points of the real line $\mathbb{R}$, because the denominators sometimes become zero. There are two natural ways to mend this situation. One is to exclude the two points 0 and 1 from $\mathbb{R}$ (note that these are the only values of x where the denominators may vanish). The other is to add one extra point ∞ to $\mathbb{R}$ and extend the functions to this point according to the rule

	F_1	F_2	F_3	F_4	F_5	F_6
0	∞	1	0	1	∞	0
1	1	0	1	∞	0	∞
∞	0	∞	∞	0	1	1

This table defines a genuine action of the group Φ on the set $\mathbb{R} \cup \infty$. It shows, in particular, that the three points 0, 1 and ∞ form one orbit, and that the action of Φ on this orbit defines the isomorphism of Φ with the permutation group on three symbols S_3.

Exercise 109. Find one more 3-element orbit of this action, and prove that all the remaining orbits consist of 6 elements.

Exercise 110. Functions $F_i \in \Phi$ can also be considered as functions of a *complex* variable. Therefore, the action of the group Φ can be prolonged to the set $\mathbb{C} \cup \infty$. Find a 2-element orbit of this action and prove that all the other orbits, except this one and the two 3-element orbits indicated above, consist of 6 elements.

5. Enumeration of orbits

Consider one more example of a group action.

Problem 44. *Let Q be a cube in space and G the group of all space rotations that take Q into itself (i.e., the proper symmetry group of Q). Make up the list of all elements of Q and describe the action of this group on the set of faces of the cube.*

Solution. Apart from the identity, the group G contains:

- 6 rotations through $180°$ around the lines that go through the midpoints of two parallel edges (like AA' in Figure 3),
- 3 rotations through $180°$ around the lines connecting the midpoints of two opposite faces (like BB'),
- 6 rotations through $90°$ around the same lines, and
- 8 rotations through $120°$ around the lines that contain a pair of opposite vertices of the cube (like CC').

The group G thus consists of 24 elements.

The natural action of the group G on the set F of faces of the cube is *transitive*, i.e., any face can be taken into any other by a suitable element of the group. In other words, the set F, which consists of 6 points, makes one orbit of the group action. For any face $f \in F$ there are exactly 4 rotations that preserve it: the identity and the three rotations around the line passing through the midpoint of this face. Note that $6 \cdot 4$ gives 24, the order of the group.

Exercise 111. Describe the action of G on the set E of edges and on the set V of vertices.

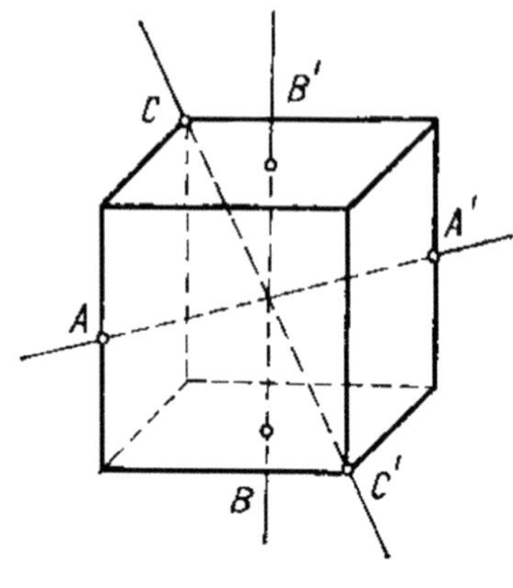

Figure 3. Rotations of the cube

So far, we have found transitive actions of the group G on the sets of order 6, 8 and 12. Note that all these numbers are divisors of 24, the order of the group. The number 24 has some more divisors, and it turns out that for every divisor one can construct a set of corresponding cardinality that consists of certain geometric elements of the cube and on which our group acts transitively. Figure 4 shows the sets of 4, 3 and 2 elements endowed with a natural transitive action of our group: these are the set D of big diagonals, the set M of middle lines and the set T of regular tetrahedra inscribed in the cube.

Definition 23. A set with a transitive action of a group is called a homogeneous space of this group.

In each of these cases we have a homomorphism of the group G into the group of transformations of the corresponding set.

Exercise 112. For which of the sets F, E, V, D, M, T is this homomorphism (a) an epimorphism? (b) an isomorphism?

We can also consider the action of the group G on more complicated objects that consist of several elements of the above sets.

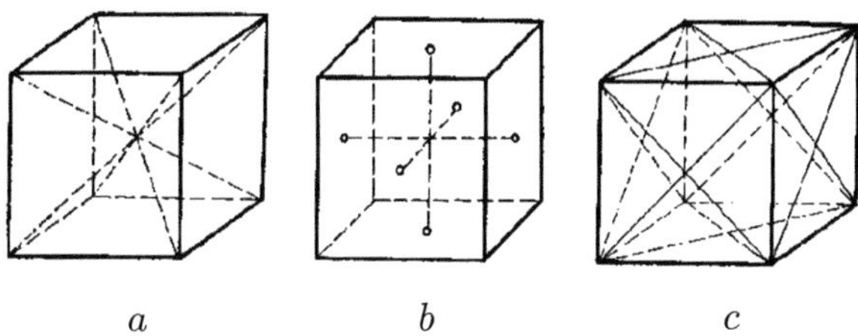

Figure 4. Homogeneous spaces of the symmetry group of the cube

Problem 45. *Describe the orbits of the action of our group G on the set of all pairs (f, e), where $f \in F$ is an arbitrary face and $e \in E$ is an arbitrary edge of the cube Q.*

Solution. The set in question can be denoted as $F \times E$ (the Cartesian product of the sets F and E). It consists in all of $6 \cdot 12 = 72$ elements which come in three different categories: the edge e can either belong to the face f (as in Figure 5a), have one common vertex with f (as in Figure 5b) or, finally, have no common points with f (Figure 5c).

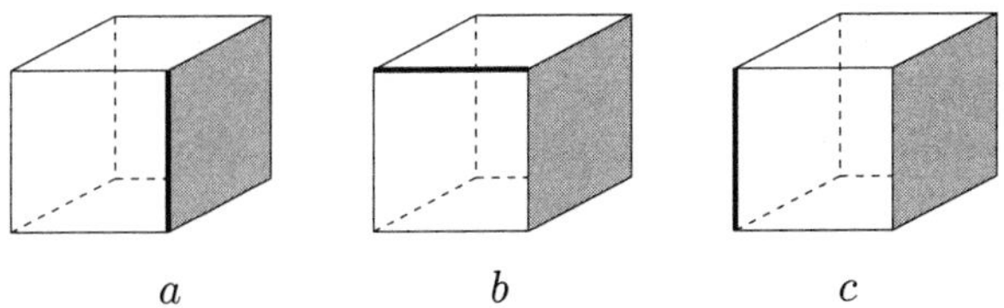

Figure 5. Different face–edge pairs

It is clear that an edge–face pair belonging to one of these types cannot go into a pair of another type under any movement. Let us prove that any two pairs of the same type can be transformed one into another by a suitable movement. Suppose that we are given two pairs

(f_1, e_1), (f_2, e_2) of the first type (edge lies in the face). First we find a rotation which takes f_1 into f_2. Then, using the 4 rotations that preserve this face, we can move e_1 to the position of e_2, and that's all. The two other cases are treated similarly.

We have thus proved that the set $F \times E$ with the action of the group G consists of three orbits, typical representatives of which are shown in Figure 5.

Exercise 113. Find the number of orbits and indicate their representatives for the action of G on the following sets: (a) $V \times F$ (vertex–face pairs), (b) $D \times F$ (diagonal–face pairs), (c) $E \times E$ (ordered pairs of edges).

Let us now generalize the observations that we have made in the previous discussion. To do so, we will need the notion of the *stabilizer*, or *stable subgroup*, of a point.

Definition 24. Given an action of a group G on the set X, the *stabilizer* of a point $x \in X$ is the set of all elements of G that preserve the point x:

$$\mathrm{St}(x) = \{g \in G \mid T_g(x) = x\}.$$

The stabilizer is a subgroup in G. Indeed, if both g and h are in $\mathrm{St}(x)$, then we have $T_{gh}(x) = T_g(T_h(x)) = T_g(x) = x$ and $T_{g^{-1}}(x) = T_g^{-1}(x) = x$.

The stabilizer of a fixed point coincides with the whole group G. For example, look at Figure 1. The stabilizer of O is the whole group, the stabilizer of the point A consists of two elements (identity and reflection), while the stabilizer of the point B is trivial (contains only the identity).

You can here notice the same rule that we saw in Problem 44: the order of the stable subgroup multiplied by the length of the corresponding orbit always gives the order of the whole group:

$$|\mathcal{O}(x)| \cdot |\mathrm{St}(x)| = |G|. \tag{37}$$

To prove this fact, consider the left coset decomposition of the group G with respect to the subgroup $H = \mathrm{St}(x)$:

$$G = g_1 H \cup g_2 H \cup \cdots \cup g_k H.$$

All the elements of the same coset act on x in the same way: if $h \in H$, then $T_{gh}(x) = T_g(T_h(x)) = T_g(x)$, which does not depend on the specific choice of h. Conversely, if two elements $g, k \in G$ move the point x to the same position, then they belong to one and the same coset. Indeed, there is an element $h \in G$ such that $k = gh$. Then

$$T_h(x) = T_{g^{-1}k}(x) = T_g^{-1} T_k(x) = x,$$

which means that $h \in \mathrm{St}(x)$. Therefore, the number of different points in the orbit of x is the same as the number of cosets in the decomposition $G/\mathrm{St}(x)$, and the assertion follows.

Formula (37) shows, in particular, that the stabilizers of all the points that belong to the same orbit have the same number of elements. In fact, these subgroups are always conjugate to each other in the group G, and hence isomorphic.

To prove this fact, take two arbitrary points x and y that belong to one and the same orbit. Then there is a group element g such that $y = T_g(x)$. We claim that the subgroups $\mathrm{St}(x)$ and $\mathrm{St}(y)$ are conjugate by the element g. Indeed, if $h \in \mathrm{St}(x)$, then

$$T_{ghg^{-1}}(y) = T_g(T_h(T_g^{-1}(y))) = T_g(T_h(x)) = T_g(x) = y,$$

which means that $ghg^{-1} \in \mathrm{St}(y)$; and, since h is arbitrary, this implies that $g\,\mathrm{St}(x)g^{-1} \subset \mathrm{St}(y)$. Interchanging the roles of x and y in the previous argument proves the inverse inclusion. Thus $g\,\mathrm{St}(x)g^{-1} = \mathrm{St}(y)$, and the two subgroups are indeed conjugate.

Problem 46. *How many different ways are there to paint the disk divided into p equal parts using n colours? The number p is assumed to be prime. Two colourings are considered to be the same, if one of them goes into another by a rotation of the disk.*

Solution. We deal with the action of the cyclic group C_p on the set of all possible n^p colourings of the disk, and we are asked to find the number of orbits of this action. The length of the orbit, which is always a divisor of the order of the group, in this example can take only two values: 1 and p, because p is prime. An orbit of length 1 corresponds to a colouring which is invariant under all rotations, i.e. a colouring where the whole disk is coloured with one colour. The total number of such colourings is n. The remaining colourings, whose total number is $n^p - n$, split into orbits of cardinality p, and the number of the orbits is $(n^p - n)/p$. The total number of orbits, i.e., the number of different ways to colour the disk, is thus $(n^p - n)/p + n$.

Note that we have proved, as a byproduct, that the number $n^p - n$ is always divisible by p, if p is prime. This is yet another proof of Fermat's little theorem (34).

Exercise 114. Try to solve the same problem in the case when p is not necessarily prime.

It is rather difficult to solve this exercise by a direct argument, like the one we used in the previous problem. However, there exists a general formula that computes the number of orbits for any group action — the so-called *Burnside formula*, which we shall now state and prove.

If $g \in G$ is an element of the group acting on a set X, then we denote by $N(g)$ the number of fixed points of the corresponding transformation T_g, i.e. the number of points $x \in M$ such that $T_g(x) = x$.

Theorem 11 (Burnside's formula). *The number of orbits r is "the arithmetic mean of the number of fixed points for all elements of the group":*

$$r = \frac{1}{|G|} \sum_{g \in G} N(g),$$

or, in other words, the sum of numbers $N(g)$ for all the elements of the group is $|G|$ times the number of orbits.

Proof. To prove the formula, let us think for a while about the following question: *how many times does a given point $x \in M$ participate in the total sum $\sum_{g \in G} N(g)$?*. Evidently, it comes up every time when an element g preserves x. The answer to the question is thus $|\operatorname{St}(x)|$. Other points belonging to the orbit $\mathcal{O}(x)$ appear in the total sum the same number of times, because all the stabilizers have the same cardinality. Therefore, the contribution of this orbit is $|\operatorname{St}(x)| \cdot |\mathcal{O}(x)|$, which, as we know, is equal to $|G|$, the order of the group. Since every orbit gives the same contribution $|G|$, the whole sum is $|G|$ times the number of the orbits, and this is exactly what we wanted to prove. □

Let us use Burnside's formula to solve Problem 46 once again. Here the identity transformation has n^p fixed points, while every non-identity transformation has n fixed points. Therefore, the number of orbits is $r = (n^p + (p-1)n)/p$. It is easy to see that this result does not differ from the one obtained before.

Problem 47. *Find the number of different necklaces made of 7 white and 3 black beads.*

Solution. It is natural to treat two necklaces as equal, if they differ only by a rotation or a reflection of the circle. Therefore, we have to consider the set X whose elements are all possible dispositions of 7 white and 3 black beads in the vertices of a fixed regular 10-gon and the action of the dihedral group D_{10} on X. The problem is to find the number of orbits of this action.

Having Burnside's formula in mind, let us calculate the number of fixed points in X for every element of the group D_{10}. For the identity transformation, all $\binom{10}{3} = 120$ points of the set X are fixed.

A nontrivial rotation cannot leave any necklace invariant. The same is true for the reflections of type (a) (see Figure 6), because the number of beads of each colour is odd. But the reflections of type (b) do have invariant necklaces. For every such reflection there are $2 \cdot 4 = 8$ such necklaces: first, one of the beads on the symmetry axes must be white and another one black, second, one of the four symmetrical pairs of beads must be black. The number of reflections of type (b) is 5, and by Burnside's formula we get $r = (120 + 5 \cdot 8)/20 = 8$.

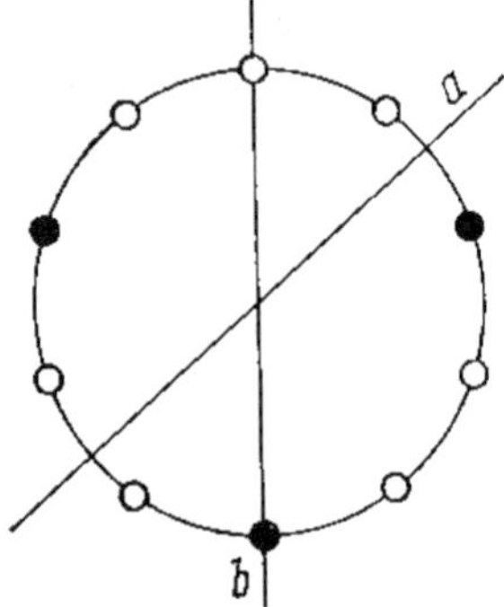

Figure 6. A necklace of type 3-7

Exercise 115. Under the central symmetry (rotation through 180°) digits 0, 1 and 8 are preserved, digits 6 and 9 change places, while all the remaining digits lose their meaning. How many 5-digit numbers are centrally symmetric?

Exercise 116. Solve the last problem in the case of 6 white and 4 black beads.

Exercise 117. A *dice* is a cube marked on each side with numbers 1 through 6. How many different dice are there? (Two dice are regarded as the same, if they can be so rotated in space that the numbers on corresponding sides become equal.)

Exercise 118. How many different ways are there to paint (a) the vertices, (b) the edges of a cube by two colours (i.e. using no more than two given colours)? (Here, as in the previous exercise, only proper rotations should be considered.)

Exercise 119. How many different hexagons can be inscribed into a regular 15-gon? (Two figures are equal, if they coincide after a plane movement, not necessarily proper.)

6. Invariants

The problem about necklaces can be solved by a simple direct argument, without referring to group theory and Burnside's formula. The most natural solution can be stated as follows.

The three black beads break the circle into three parts, which contain m, n and k white beads. The integers m, n and k are between 0 and 7, inclusive, and satisfy $m + n + k = 7$. Note that the order in which these three numbers appear is in our case irrelevant, because rotations and reflections produce any of the 6 possible permutations. (For the case of 4 black beads this observation would no longer be true!) Therefore, we can assume that $m \leq n \leq k$, and the problem is reduced to the enumeration of all triples of integers that satisfy all the stated restrictions. All such triples can be found directly. Here they are, in lexicographic order: $(0, 0, 7)$, $(0, 1, 6)$, $(0, 2, 5)$, $(0, 3, 4)$, $(1, 1, 5)$, $(1, 2, 4)$, $(1, 3, 3)$, $(2, 2, 3)$.

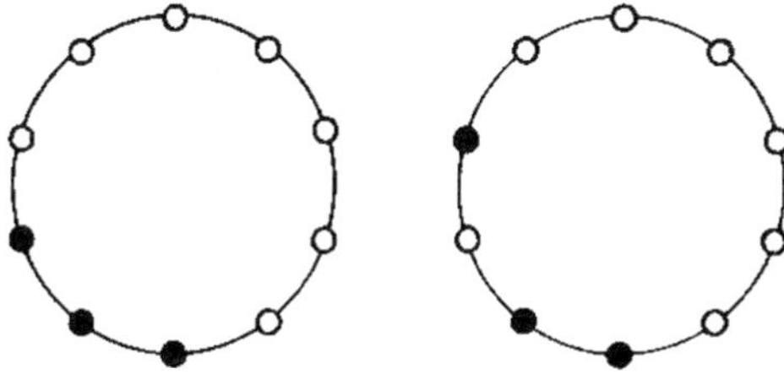

Figure 7. Invariant of a necklace

Why is it that the triple (m, n, k) can serve for the enumeration of orbits? Because it satisfies the following two properties:

- first, if two necklaces are the same (belong to the same orbit), then the corresponding triples are equal,
- second, if the triples of two necklaces are equal, then the necklaces themselves are equivalent.

The first property can also be expressed in the following way: the values of m, n and k are constant on the orbits of the action.

Consider the general situation. Suppose that the group G acts on the set X.

Definition 25. A mapping φ from X into a certain set N is called an *invariant of the action*, if the values it takes on the elements of the same orbit are always equal.

Invariants may take values in arbitrary sets. In the previous example, the set N consisted of unordered triples of integers. The least of these three numbers (denoted above by m) is also an invariant of the group action under study. However, this invariant does not possess the second property: for example, the two necklaces shown in Figure 7 are different, but the values of m for them are the same. Such an invariant cannot be used to distinguish different orbits.

Definition 26. An invariant $\varphi : X \to N$ is said to be *complete*, if it takes different values on different orbits.

> **Exercise 120.** Construct a complete invariant for necklaces with 4 black and 6 white beads.

Let $\mathcal{M}$ be the group of all plane movements. This group acts on the plane in a natural way. This action is transitive; thus the plane is a homogeneous space of the group $\mathcal{M}$. Invariants of this action present no interest: these are only constant mappings.

The group G also acts on the set of all straight-line segments in the plane. This action is not trivial. Two segments belong to one orbit, if and only if their lengths are equal. The length of the segment is thus a complete invariant of this action.

> **Exercise 121.** Indicate some complete invariants for the action of the group G of plane movements on the set of (a) all triangles, (b) all quadrangles.

If $H \subset G$ is a subgroup, then it also acts on the plane. If H is small enough, then its orbits cannot be big, and therefore the action might have nontrivial invariants. For example, if H is the group of rotations around a point A, then its orbits are circles centred at A. The distance of a point from A is the complete invariant of this action. In the polar system of coordinates with centre A (see p. 33), the distance is the polar coordinate r, and every invariant has the form $f(r)$, i.e., is a function of the complete invariant.

Consider the action of the dihedral group D_3 on the plane (see Figure 1). Let O be the polar centre and OM the polar axis. The polar distance r is an invariant of this action. But it is not a complete invariant. To make it complete, we will add one more function to r. Note that the function $\cos\varphi$ is an invariant of the group D_3.

Indeed, the group is generated by the reflection $\varphi \mapsto -\varphi$ and the rotation $\varphi \mapsto \varphi + 120°$, and the expression $\cos 3\varphi$ does not change under these transformations. The pair of numbers $(r, \cos 3\varphi)$ makes a complete invariant of the action. Indeed, it is easy to check that the simultaneous equations

$$\cos 3\varphi = b,$$
$$r = c$$

for any real $c > 0$ and $|b| \leq 1$ may have 3 or 6 solutions that correspond to the points of one orbit.

7. Crystallographic groups

We now have all the techniques necessary to revisit the question about the symmetry of ornaments posed in the introduction (see page 4). The symmetry of ornaments — plane patterns infinitely repeated in two or more different directions — is described by the so-called *plane crystallographic groups.* An example of such a group is the rolling group of the equilateral triangle studied in Problem 27 — it describes the symmetry of the ornament shown in Figure 4b in the Introduction. Crystallographic groups are also referred to as *wallpaper groups.*

The exact definition reads as follows.

Definition 27. A *crystallographic group* is a discrete group of plane movements that has a bounded fundamental domain.

We will explain the two terms that appear in this definition.

Definition 28. A group of plane movements G is said to be *discrete,* if every orbit is a discrete set in the plane, i.e., for every point A there is a disk centred at A and containing no other point of the same orbit.

A simple example of a discrete group is provided by the cyclic group generated by one translation. On the contrary, the group containing two translations with collinear vectors of incommensurable lengths is not discrete, because the orbit of every point A is a dense subset of the line passing through A in the direction of the translations.

Exercise 122. Prove that the group generated by a rotation through α degrees is discrete if and only if the number α is rational.

Exercise 123. Prove that the stabilizer of any point with respect to a discrete group of plane movements is finite.

The second notion that needs explanation is that of a *fundamental domain.*

Definition 29. A domain[1] F is said to be *fundamental for the group* G, if

- any point in the plane belongs to the orbit of some point $x \in F$ (which can also be a boundary point), and
- no two different inner points of F belong to the same orbit.

These two properties mean that the images of the domain F under the group transformations are all distinct (with the exception of boundary points) and fill the plane without overlapping. Another wording is that we have a *tiling*, or *tessellation*, of the plane by copies of the figure F. For example, the rolling group of the equilateral triangle (Problem 27) is crystallographic, and the initial triangle can be chosen as its fundamental domain. The assertion claimed in the statement of this problem is exactly the second property in the definition of a fundamental domain.

Exercise 124. Find the fundamental domains for the groups C_n and D_n.

The term "crystallographic" has its origin in the fact that discrete groups of *space* movements are used to describe the symmetry of natural crystals. There exists a special universal system of notation for the crystallographic groups, both plane and spatial. For example, the rolling group of the equilateral triangle is traditionally denoted by $p3m1$. We will give some more examples of crystallographic groups and corresponding ornaments.

The simplest such group, denoted by $p1$, is the group generated by two translations by non-collinear vectors **a** and **b**. Figure 8a shows the generators of the group and the orbit of a point.

---

[1]It would take some effort to give an exact meaning of the notion of domain. However, it is safe to replace the word "domain" by "polygon" everywhere in this section.

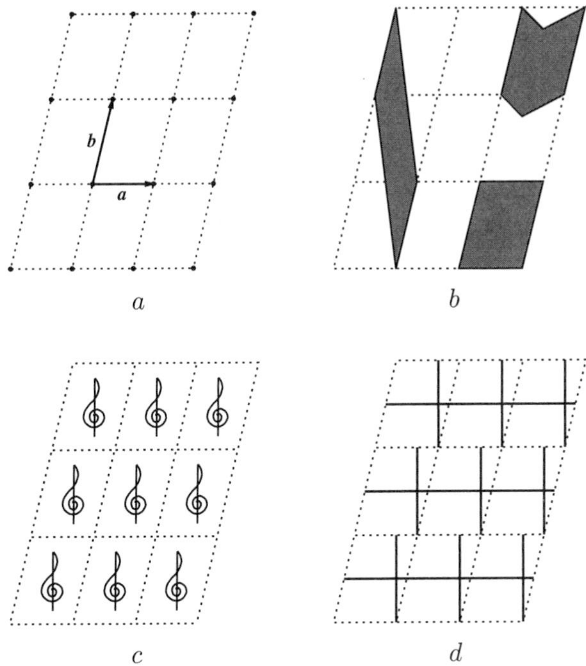

Figure 8. The simplest crystallographic group

As a fundamental domain, one can take the parallelogram with sides **a** and **b**. The fundamental domain can be chosen in a variety of ways. Figure 8b shows several different fundamental domains for the same group $p1$. Two of these domains are parallelograms, while the third one is a hexagon. The sides of the parallelogram are $\mathbf{b} - \mathbf{a}, \mathbf{b}$ and $\mathbf{b}$, $2\mathbf{b} - \mathbf{a}$, respectively.

Exercise 125. Prove that the parallelogram with sides $k\mathbf{a} + l\mathbf{b}$ and $m\mathbf{a} + n\mathbf{b}$, where k, l, m, n are integers, is a fundamental domain if and only if $|kn - lm| = 1$.

The group $p1$, simplest among all ornamental groups, describes the purely translational symmetry of an ornament. An ornament has symmetry $p1$, if it has no symmetries other than translations. It is very easy to invent such an ornament. All you have to do is draw an arbitrary figure with a trivial symmetry group, lying inside the fundamental parallelogram, and consider the union of all the copies of this figure obtained by parallel translations of the given group (Figure 8c).

If the chosen figure lies strictly inside the parallelogram, then the ornament obtained has symmetry group exactly equal to $p1$. However, if you allow the figure to touch the border, then the ornament may have a wider symmetry group. An example of this phenomenon is shown in Figure 8d.

The group $p1$ is important not only because it is the simplest crystallographic group in the plane, but also because of the following fact.

Lemma 1. *Every plane crystallographic group contains a subgroup of type* p1, *i.e., generated by two non-collinear translations.*

Proof. A crystallographic group must contain at least one translation, as otherwise it will reduce to a finite group.

Suppose that G is a discrete group of plane movements such that all the translations belonging to G have the same direction (say, parallel to the line l). We are going to prove that the fundamental domain of the group G cannot be bounded in this case.

Let us first consider what movements, other than rotations, can belong to the group G. Note first that the axes of all glide symmetries that belong to G must also be parallel to the line l — because the square of a glide symmetry is a translation. The rotations that belong to the group can only be rotations through 180°, because, if $R_\varphi \in G$ and $\varphi \neq 180°$, then, together with every translation $T_\mathbf{a} \in G$ the group also contains the translation $R^\varphi \circ T_\mathbf{a} \circ R^{-\varphi}$, non-collinear with $T_\mathbf{a}$ (see the answer to Exercise 58). The composition of two rotations by 180° is a translation along the line that connects their centres. Therefore, the centres of all rotations that belong to G must lie on one line parallel to l. Without loss of generality, we can assume that l is the line that passes through the centres of all rotations. Finally, we leave it to the reader as an exercise to find out what kind of reflections our group may contain.

Exercise 126. Prove that the group G may contain only reflections whose axes are perpendicular to l or coincide with l.

From all the observations made, we can conclude that two points can belong to one orbit only if their distances from the line l are equal. The fundamental domain must contain one point of each orbit; therefore it cannot be bounded, and the group G is non-crystallographic. □

We have thus proved that every ornamental group G contains two non-collinear translations, and hence a subgroup of type $p1$ that they generate. In fact, a stronger assertion holds.

Lemma 2. *The set of all translations belonging to G is a group of type $p1$.*

This lemma is an immediate consequence of the following exercise.

Exercise 127. Prove that every ornamental group that consists only of translations is generated by two non-collinear translations.

In a certain sense, any ornamental group is reduced to the group $p1$ and a finite group of plane movements. We will explain how. Let G be an arbitrary ornamental group. Denote by H its subgroup of translations (we already know that it is of type $p1$). The subgroup H is normal in G, because the conjugate of a translation by any movement is always a translation (see page 234).

Lemma 3. *For any crystallographic group G, the quotient group G/H is finite and may only belong to one of the ten types C_n, D_n, where $n = 1, 2, 3, 4, 6$.*

We are not going to prove this fact now. The reader will verify it later, using the table of plane crystallographic groups (Exercise 130). The type of the group G/H is referred to as the *ornamental class* of G.

Let us now discuss the relation between the groups G and H and their fundamental domains in more detail.

If Φ is a fundamental domain of G and g_1, ..., g_k is the complete set of representatives of the coset decomposition G/H, then the union

$$\Pi = \bigcup_{i=1}^{k} g_i \Phi$$

forms a fundamental domain for H. The domain Φ is called the *motif* of the ornament, and the domain Π, its *elementary cell*. In principle, Φ and g_i can always be chosen in such a way that Π becomes a parallelogram, but sometimes it is more convenient to use polygons of another shape, notably, regular hexagons, as the elementary cell.

The ratio of the area of Π to the area of Φ is equal to the index of H in G, i.e., the number of cosets in G/H. The bigger G/H, the smaller the fundamental domain Φ in Π.

Problem 48. *Find the motif and the elementary cell of the ornament that has symmetry group of type* p3m1. *Describe the cosets of* G *with respect to its subgroup of translations* H.

Solution.

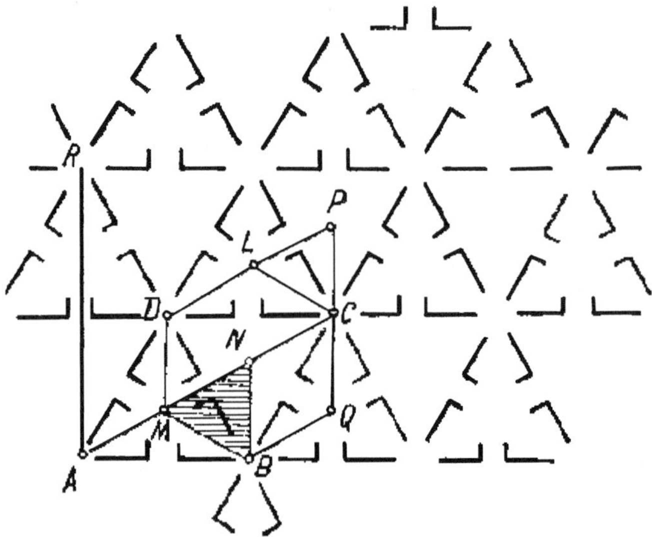

Figure 9. An ornament with symmetry group $p3m1$

Look at Figure 9a, which shows an ornament with symmetry group $p3m1$, i.e., the group generated by three reflections in the sides of an equilateral triangle. (Instead of Figure 9a, we could as well use Figure 4b from the Introduction.) The ornament is obtained by rolling over the triangle MNB (the *motif*) on the plane. You can see from the picture that the ornament admits translations by vectors that connect the points A, B and D. As the elementary cell we can choose, for example, the parallelogram $ABCD$, whose sides $\overrightarrow{AB}$ and $\overrightarrow{AD}$ generate the subgroup of translations H. It is easy to see that $S_{ABCD} : S_{MNB} = 6 : 1$. It is impossible to divide the parallelogram $ABCD$ into six equilateral triangles which are the fundamental domains of the ornament (although it is possible to divide it into six fundamental domains of another type — try to do this!). In this case, it is more convenient to choose the fundamental domain for H having the shape of a hexagon, for example $BQCLDM$, which is naturally divided into six fundamental triangles.

To find the number of cosets in G/H, let us note that the elements of the subgroup H do not change the relative position of the motif in the plane (the direction and the orientation of the "leg"). Different movements that belong to the same coset, say $g \circ h_1$ and $g \circ h_2$, $h_1, h_2 \in H$, change the relative position of the motif in the same way, because h_1 and h_2 do not change it at all. In the picture, you can see six different positions of the motif; hence the number of the cosets in G/H is six. We can number them from 1 to 6, as in Figure 9b. Then, for example, all the movements belonging to the coset gH, where $g \in G$ is the reflection in a vertical line, induce the following permutation of these numbers: $1 \leftrightarrow 6$, $2 \leftrightarrow 5$, $3 \leftrightarrow 4$.

We have thus arrived at the following conclusion: the quotient group G/H acts on the set of motifs contained in the elementary cell in the same way as the group D_3 acts on the set of vertices of an equilateral triangle: there are three rotations, including the identity, and three reflections. We can write the result as follows: $G/H \cong D_3$. In the terminology introduced above, this fact can also be stated as follows: the ornamental class of G is D_3.

Note, finally, that writing $p3m1/p1 \cong D_3$ is not correct, because there are many subgroups of type $p1$ contained in the group of type $p3m1$.

Exercise 128. What is the order of the quotient group G/K, where G is the group just studied, and $K \subset H$ is the subgroup of translations generated by $\overrightarrow{AC}$ and $\overrightarrow{AR}$ (see Figure 9a). Describe the structure of this group; in particular, find whether it is isomorphic to one of the groups C_n, D_n.

How many different subgroups of types $p1$ are there in the group of plane movements $\mathcal{M}$? Of course, an infinite number: the choice of a specific group is determined by two basic vectors $\mathbf{a}$ and $\mathbf{b}$. However, any two such groups are isomorphic. In fact, a stronger assertion holds: any two such groups are conjugate to each other via a suitable linear transformation (see page 168) of the plane: if H is generated by translations $T_{\mathbf{a}}$ and $T_{\mathbf{b}}$ and K by translations $T_{\mathbf{c}}$ and $T_{\mathbf{d}}$, then $LHL^{-1} = K$, where L is a linear transformation such that $L(\mathbf{a}) = \mathbf{c}$ and $L(\mathbf{b}) = \mathbf{d}$.

Definition 30. Two groups of plane movements are said to be *equivalent*, if they are conjugate to each other by a suitable linear transformation.

We can now state the theorem that gives the exact meaning to the statement that there are 17 types of wallpaper symmetry.

Theorem 12 (Fedorov–Schoenflies). *Up to the equivalence formulated above, any plane crystallographic group is equivalent to one of the 17 groups given in the table below. These 17 groups are not isomorphic to each other.*

Outline of the proof. The proof is not very difficult, but rather lengthy. We are only giving an outline, leaving the details to the industrious reader.

(1) Prove that the only possible rotations in a crystallographic group are of order 2, 3, 4 or 6.

(2) Let $G^+ \subset G$ be the subgroup of all proper (orientation preserving) movements in the group G. Then G^+ is a normal subgroup of index 1 or 2.

(3) If $G^+ = G$, i.e. the group consists only of translations and rotations, then it is equivalent to one of the groups $p1$, $p2$, $p3$, $p4$, $p6$, depending on the biggest order of a rotation it contains.

(4) If $G^+ \neq G$, then G is generated by the subgroup G^+, which belongs to one of the five types listed above, and one movement f from $G \setminus G^+$. Considering the various possibilities that may arise (f is either a reflection or a glide reflection, its axis may pass or not pass through the centres of rotations etc.), we establish that
   (a) If $G^+ = p1$, then $G = pm$, pg or cm.
   (b) If $G^+ = p2$, then $G = pmm$, pmg, pgg or cmm.
   (c) If $G^+ = p3$, then $G = p31m$ or $p3m1$.
   (d) If $G^+ = p4$, then $G = p4m$ or $p4g$.
   (e) If $G^+ = p6$, then $G = p6m$.

□

Now, the table. For every group, the table of plane crystallographic groups includes (left to right):

- *Smbl*: The canonical crystallographic notation of the group.
- *Symmetries*: An elementary cell (either a square or a regular hexagon) with the symbols for the movements contained in the group (a solid line means the axis of a reflection, a dashed line means the axis of a glide symmetry, and the symbols ○, △, ◇, ⬣ designate the centres of rotation of orders 2, 3, 4 and 6).
- *Sample*: A sample ornament with this symmetry group. The sample shows only one cell of the ornament. The ornament is obtained from the elementary cell by translations in two

non-collinear directions. Inside the elementary cell, a fundamental domain is hatched.

- *Generators and relations*: A set of generators and defining relations of the group.

In the table, for groups number 1–12, we give a representative with an elementary cell in the form of a square, and for groups 13–17, in the form of a regular hexagon. Note that the cell can be an arbitrary parallelogram for groups $p1$ and $p2$, an arbitrary rectangle for groups pm, pg, pmm, pmg, and pgg, and an arbitrary rhombus for groups cm, cmm.

Using this table, the reader can determine the symmetry type of any ornament, starting from the wallpaper design on the walls of his or her room.

Table of plane crystallographic groups

Smbl	*Symmetries*	*Sample*	*Generators and relations*
p1		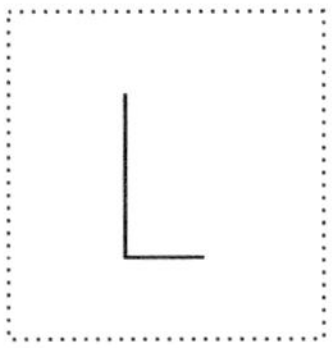	Non-collinear translations T_1, T_2 $T_1T_2 = T_2T_1$
p2	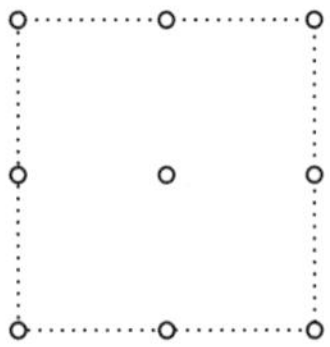	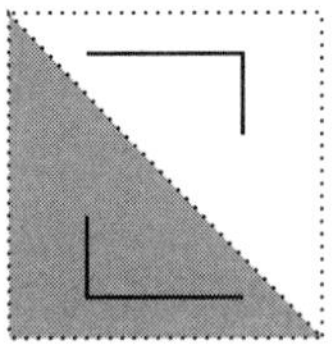	Half turns R_1, R_2, R_3 $R_1^2 = R_2^2 = R_3^3 = \mathrm{id}$, $(R_1R_2R_3)^2 = \mathrm{id}$
pm		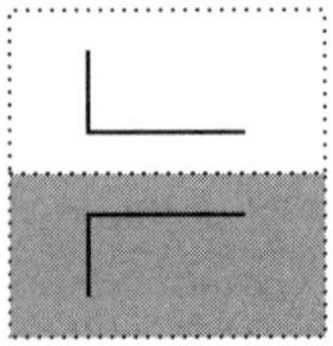	Reflections S_1, S_2 and translation T $S_1T = TS_1$, $S_2T = TS_2$, $S_1^2 = S_2^2 = \mathrm{id}$

Smbl	*Symmetries*	*Sample*	*Generators and relations*
pg		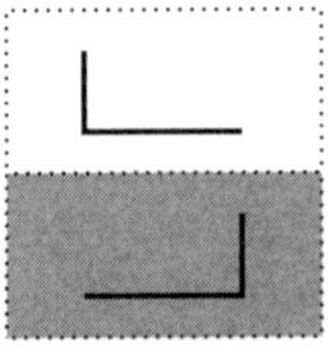	Parallel glide reflections U_1, U_2 $U_1^2 = U_2^2$
cm	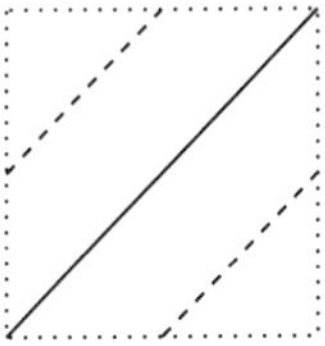	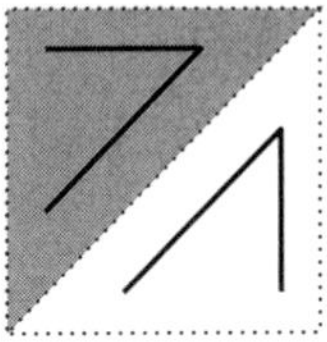	Reflection S and glide reflection U $S^2 = \text{id}$, $SU^2 = U^2S$
pmm	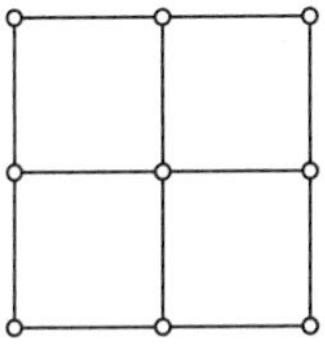	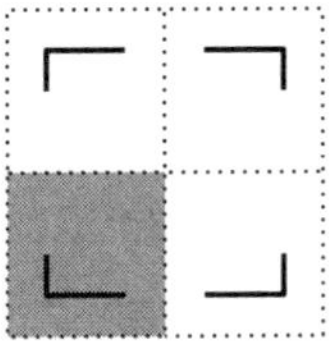	Reflections in the sides of a rectangle S_1, S_2, S_3, S_4 $S_1^2 = S_2^2 = S_3^2 = S_4^2 = \text{id}$, $(S_1S_2)^2 = (S_2S_3)^2 = (S_3S_4)^2 = (S_4S_1)^2 = \text{id}$
pmg	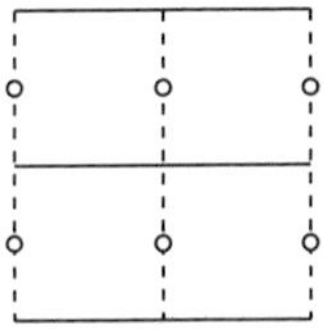	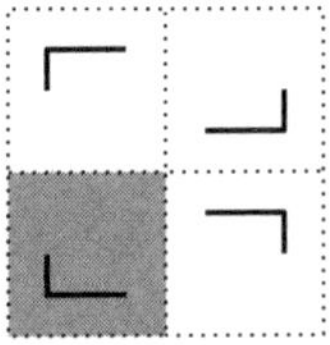	Reflection S and central symmetries R_1, R_2 $S^2 = R_1^2 = R_2^2 = \text{id}$, $R_1SR_1 = R_2SR_2$
pgg	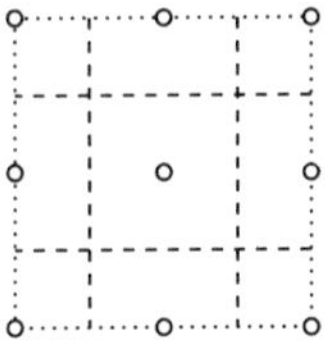	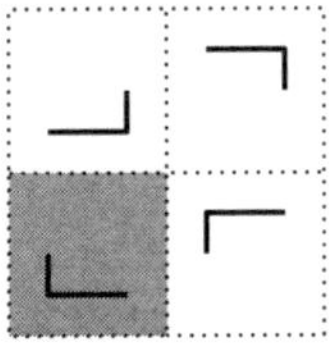	Perpendicular glide reflections U_1, U_2 $(U_1U_2)^2 = (U_1^{-1}U_2)^2 = \text{id}$
cmm	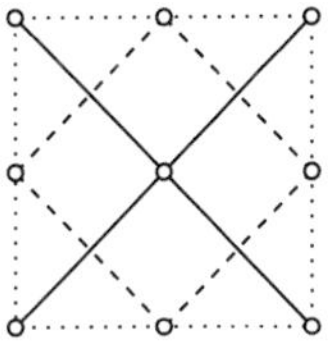	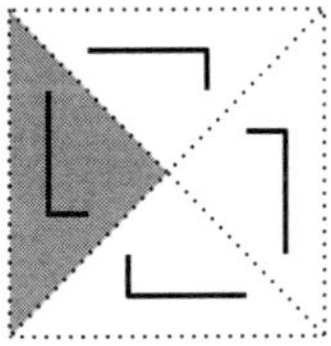	Reflections S_1, S_2 and central symmetry R $S_1^2 = S_2^2 = R^2 = \text{id}$, $(S_1S_2)^2 = (S_1RS_2R)^2 = \text{id}$

Smbl	*Symmetries*	*Sample*	*Generators and relations*
p4	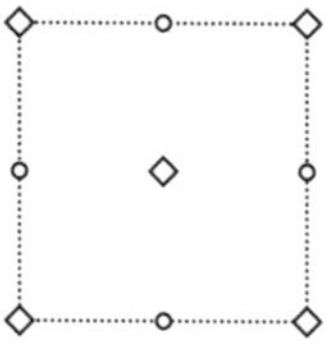	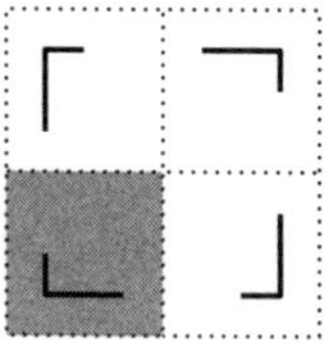	Central symmetry R and 90° rotation R_1 $R^2 = R_1^4 = (R_1R)^4 = \mathrm{id}$
p4m	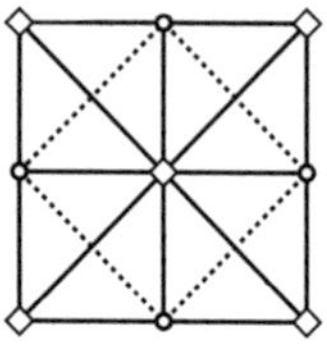	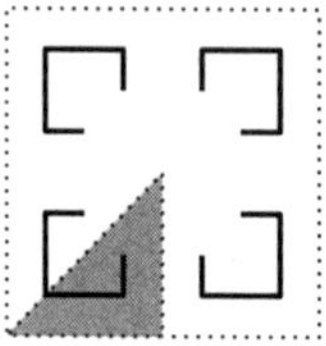	Reflections S_1, S_2, S_3 in the sides of an isosceles right triangle $S_1^2 = S_2^2 = S_3^2 = \mathrm{id}$, $(S_1S_2)^2 = (S_2S_3)^4 = (S_3S_1)^4 = \mathrm{id}$
p4g	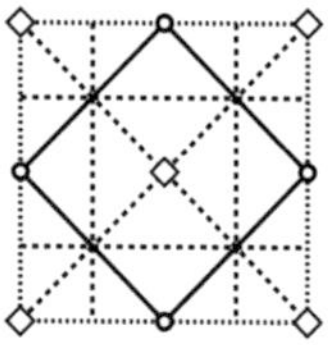	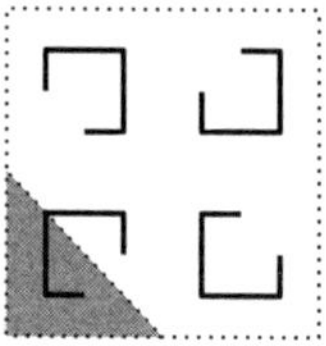	Reflection S and 90° rotation R $S^2{=}R^4{=}(R^{-1}SRS)^2{=}\mathrm{id}$
p3	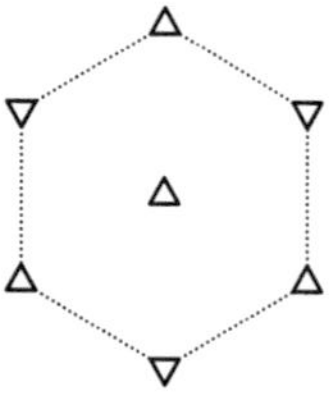	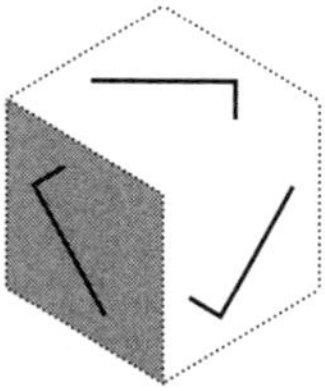	Three rotations R_1, R_2, R_3 through 120° $R_1^3 = R_2^3 = R_3^3 = R_1R_2R_3 = \mathrm{id}$
p31m	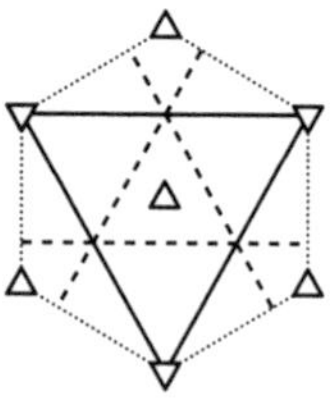	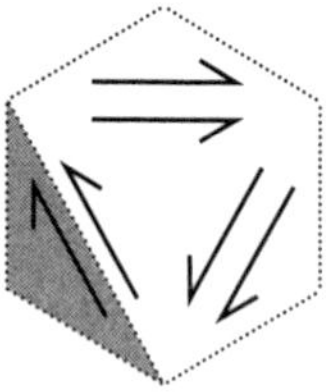	Reflection S and rotation R through 120° $R^3 = S^2 = (R^{-1}SRS)^3 = \mathrm{id}$
p3m1	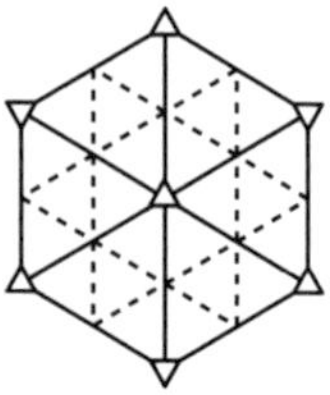	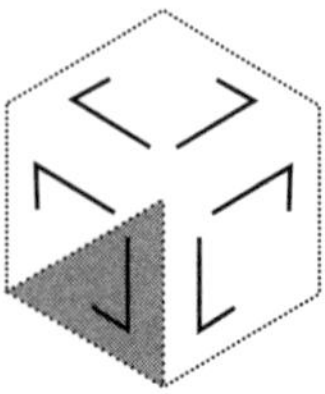	Reflections S_1, S_2, S_3 in the sides of an equilateral triangle $S_1^2 = S_2^2 = S_3^2 = \mathrm{id}$, $(S_1S_2)^3 = (S_2S_3)^3 = (S_3S_1)^3 = \mathrm{id}$

Smbl	*Symmetries*	*Sample*	*Generators and relations*
p6			Half turn R and $120°$ rotation R_1 $R^2 = R_1^3 = (R_1R)^6 = \mathrm{id}$
p6m			Reflections S_1, S_2, S_3 in the sides of a $(30°, 60°, 90°)$ triangle $S_1^2 = S_2^2 = S_3^2 = \mathrm{id}$, $(S_1S_2)^2 = (S_2S_3)^3 = (S_3S_1)^6 = \mathrm{id}$

Exercise 129. Find the symmetry groups of the ornaments shown in Figure 4 (page 4) and Figure 8d.

Exercise 130. Determine the ornamental class of every group in the table, and thus prove Lemma 3 (p. 155).

Exercise 131. Try to guess the meaning of the letters and numbers used in the notation of crystallographic groups.

Chapter 6

Other Types of Transformations

The main protagonists of the book — transformation groups — have so far appeared in the particular case of groups of plane movements. In the present chapter, we are going to discuss other types of plane transformations: affine and projective transformations, similitudes and inversions. All these transformations can be described by fractional linear functions of either two real or one complex argument.

1. Affine transformations

Affine transformations constitute an important class of plane transformations which is a natural generalization of movements. In fact, the group of plane movements $\mathcal{M}$ is a subgroup of the affine group $\mathrm{Aff}(2, \mathbb{R})$. The transition from movements to affine transformations is easily achieved in coordinates.

Problem 49. *Find a description of plane movements in Cartesian coordinates.*

Solution. If Oxy is a system of Cartesian coordinates in the plane (which we will call the 'old' system), then its image under a movement f is another Cartesian coordinate system $O_1x_1y_1$, called the 'new' system.

If a point A has coordinates (p,q) in the old coordinate system, then its image A' has the same coordinates (p,q) in the new system.

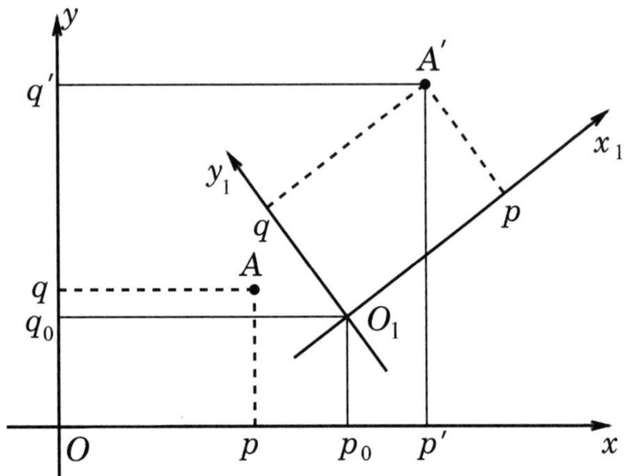

Figure 1. Movement in coordinates

For an orientation-preserving movement (translation or rotation), it is easy to find the coordinates (p',q') of point A' in the old system, using Figure 1:

(38)
$$\begin{cases} p' = p\cos\alpha - q\sin\alpha + p_0, \\ q' = p\sin\alpha + q\cos\alpha + q_0, \end{cases}$$

where α is the angle between the rays Ox and O_1x_1.

For an orientation-reversing transformation we get similar formulas with q changed to $-q$:

(39)
$$\begin{cases} p' = p\cos\alpha + q\sin\alpha + p_0, \\ q' = p\sin\alpha - q\cos\alpha + q_0. \end{cases}$$

Affine transformations are given by a formula similar to (38) and (39), where the coefficients may be arbitrary numbers, not necessarily sines and cosines.

Definition 31. An *affine transformation of the plane* is a transformation that takes a point (x, y) to the point (x', y') according to the equations

$$
\left\{ \begin{array}{lcl} x' & = & ax + by + x_0, \\ y' & = & cx + dy + y_0. \end{array} \right. \tag{40}
$$

Formulas (40) make sense for any values of the coefficients a, b, c, d. However, if we want to obtain a genuine (one-to-one) transformation of the plane, we must suppose that $ad - bc$ (the determinant of the matrix $\left(\begin{smallmatrix} a & b \\ c & d \end{smallmatrix}\right)$) is different from zero. Indeed, basic vectors $(1, 0)$ and $(0, 1)$ are taken, by the transformation (40), into vectors (a, c) and (b, d), and the area of the parallelogram constructed on these two vectors is equal to $ad - bc$.

Under an affine transformation, parallel lines go into parallel lines, but the angles are not preserved: a square may become an arbitrary parallelogram. Figure 2 shows an example of an affine transformation with the matrix $\begin{pmatrix} a & b \\ c & d \end{pmatrix}$ equal to $\begin{pmatrix} 1 & 1/2 \\ 0 & 1 \end{pmatrix}$.

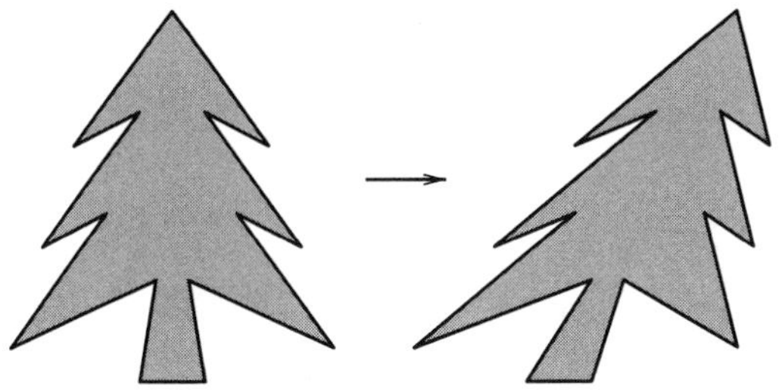

Figure 2. An affine transformation

Definition 32. The group of affine transformations of the plane, denoted by $\mathrm{Aff}(2, \mathbb{R})$, consists of all affine transformations (40) with $ad - bc \neq 0$.

The group of affine transformations acts transitively on the plane. The stable subgroup of the origin O is the *group of linear transformations* $\mathrm{GL}(2, \mathbb{R})$.

Definition 33. A *linear transformation of the plane* is a transformation given in Cartesian coordinates by the equations

$$\begin{cases} x' = ax + by, \\ y' = cx + dy. \end{cases} \tag{41}$$

The group $\mathrm{GL}(2, \mathbb{R})$ consists of all such transformations with $ad - bc \neq 0$.

Exercise 132. Let $\mathbb{R}^2$ denote the group of plane translations. Prove the isomorphism $\mathrm{Aff}(2, \mathbb{R})/\mathbb{R}^2 \cong \mathrm{GL}(2, \mathbb{R})$.

The fundamental property of affine transformations is that they preserve the ratio of points on straight lines (see page 16): if a point C divides a segment AB in the ratio $k : l$, then its image C' will divide the corresponding segment $A'B'$ in the same ratio $k : l$. In fact, one can prove that $\mathrm{Aff}(2, \mathbb{R})$ *coincides* with the set of all transformations of the plane that take straight lines into straight lines and preserve the ratio of points on every line.

Affine transformations are useful for the solution of geometric problems where the statement is invariant under affine transformations, but the solution is easier for some special case of the construction.

To take a simple example, consider the property of the medians that we talked about in Chapter 1: *the three medians in any triangle meet in one point and this point divides each of them in the ratio* $2 : 1$ (see Exercise 10). By an affine transformation, the given triangle can be reduced to an equilateral one, for which the assertion is evident.

Exercise 133. Find another solution of Problem 4 (page 19), using affine transformations.

The notions of linear and affine transformations make sense in the one-dimensional case, too. Linear transformations of the line have the form $x \mapsto ax$, while affine transformations are described by the formula $x \mapsto ax + b$. The corresponding groups are distinguished by the condition $a \neq 0$ and denoted by $\mathrm{GL}(1, \mathbb{R})$ and $\mathrm{Aff}(1, \mathbb{R})$, respectively. Instead of real numbers $\mathbb{R}$ we can consider residues over a prime number, thus arriving at finite groups. One can also use complex numbers instead of real: the corresponding groups will come up later in this chapter (see section 3).

Exercise 134. (a) How many elements are there in the group $G = \mathrm{GL}(2, \mathbb{Z}_2)$? Among the groups that we considered earlier, find a group isomorphic to G. (b) The same questions for the group $\mathrm{Aff}(1, \mathbb{Z}_3)$.

2. Projective transformations

The notion of projective transformations comes from daily life.

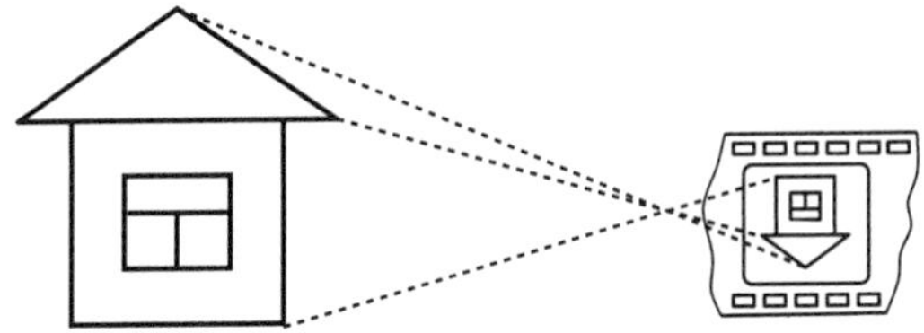

Figure 3. Photography as projection

From the mathematical point of view, photography, as well as still life drawing, is a *perspective transformation*, or a *central projection.* Photography takes every point A of the given object into the point A' where the line AO (O being the optical centre of the camera) meets the plane of the film (Figure 3). To make a drawing of nature, the artist does essentially the same thing, with the difference that the plane of the canvass is placed *between* the object and the eye.

It is clear that under such transformations straight lines go into straight lines. Hence it is possible to study the perspective transformations of a line, too.

Definition 34. Suppose that l and l' are two lines in the plane, and S is a fixed point in the same plane (Figure 4a). A *perspective transformation* is a mapping $p : l \to l'$ that takes every point $A \in l$ into the intersection point A' of the lines SA and l'. Given two planes Π_1, Π_2 and a point S in space, one can define a perspective transformation $p : \Pi_1 \to \Pi_2$ in a similar way (see Figure 4b).

Definition 35. A *projective transformation of a line or a plane into itself* is a composition of several perspective transformations where auxiliary lines or planes are used.

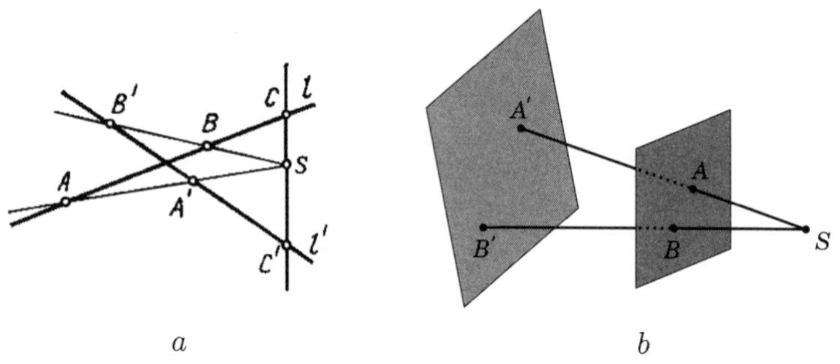

Figure 4. Perspective transformation of lines (a) and planes (b)

We begin the discussion of the properties of projective transformations with the following question. Suppose that we take a picture of a series of equidistant objects arranged along a line (e.g., trees along a road). It is clear that the images of these points on the picture do not need to be equidistant. It is also clear that they cannot be arbitrary and there must exist a certain invariant which is preserved by the projective transformations. Such an invariant must depend on more than three points, because the distance between two points can change and also the mutual relation of the three points can change arbitrarily: for example, in Figure 4a the point B lies between A and C, but its image B' is no longer between the respective images A' and C'.

The remarkable fact is that a certain function of four points, called their *cross ratio*, or *anharmonic ratio*, does not change under projective transformations of the line.

Definition 36. The *cross ratio* of points A, B, C and D is defined as follows:

$$(A, B; C, D) = \frac{AC}{BC} : \frac{AD}{BD},$$

where the lengths of the lines, AC, BC, AD, BD are considered as signed numbers, positive or negative depending on the orientation of the given pair of points.

Theorem 13. *The cross ratio of four points is preserved under projective transformations.*

Proof. It is enough to prove the invariance under perspective transformations.

To do so, let us express the areas of the triangles that you can see in Figure 5, using two different formulas:

$$S_{\triangle SAC} = \frac{1}{2}h \cdot AC = \frac{1}{2}SA \cdot SC \sin \angle ASC,$$

$$S_{\triangle SBC} = \frac{1}{2}h \cdot BC = \frac{1}{2}SB \cdot SC \sin \angle BSC,$$

$$S_{\triangle SAD} = \frac{1}{2}h \cdot AD = \frac{1}{2}SA \cdot SD \sin \angle ASD,$$

$$S_{\triangle SBD} = \frac{1}{2}h \cdot BD = \frac{1}{2}SB \cdot SD \sin \angle BSD.$$

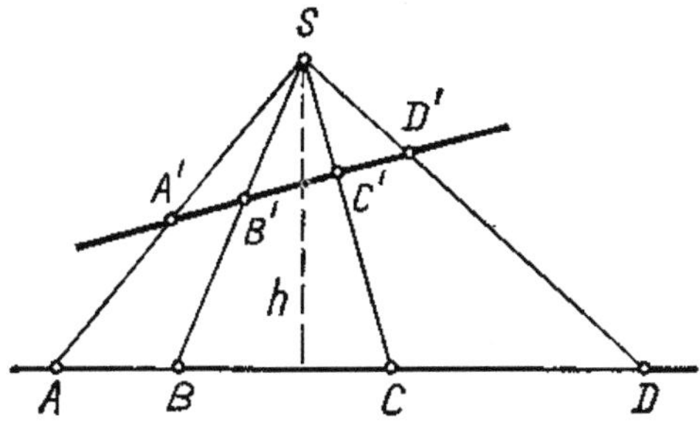

Figure 5. Derivation of the cross ratio

Now,

$$\begin{aligned}\frac{AC}{BC} : \frac{AD}{BD} &= \frac{S_{\triangle SAC}}{S_{\triangle SBC}} : \frac{S_{\triangle SAD}}{S_{\triangle SBD}} \\ &= \frac{SA \cdot SC \sin \angle ASC \cdot SB \cdot SD \sin \angle BSD}{SB \cdot SC \sin \angle BSC \cdot SA \cdot SD \sin \angle ASD} \\ &= \frac{\sin \angle ASC}{\sin \angle BSC} : \frac{\sin \angle ASD}{\sin \angle BSD}.\end{aligned}$$

We see that the cross ratio of four points is expressed through the angles at which the corresponding segments are viewed from the centre S. These angles do not change under the perspective transformation, and the theorem is proved. □

Problem 50. *Let A, B, C, D be four sequential equidistant trees along a straight-line road, and let A', B', C', D' be their images in a photo. Suppose that the distance $A'B'$ is 6 cm and the distance $B'C'$ is 2 cm. What is the distance $C'D'$?*

Solution. We have

$$\frac{AC}{BC} : \frac{AD}{BD} = \frac{2}{1} : \frac{3}{2} = \frac{4}{3}.$$

Denoting $C'D'$ by x, we have

$$\frac{A'C'}{B'C'} : \frac{A'D'}{B'D'} = \frac{8}{2} : \frac{x+8}{x+2},$$

and from the equation

$$\frac{8(x+2)}{2(x+8)} = \frac{4}{3}$$

we find that $x = 1$.

A similar argument can be used to derive the general formula that expresses projective transformations in coordinates. Suppose that x is the coordinate of a variable point M on the line l and x' the coordinate of its image $M' \in l'$. Fix three different points A, B, C on l, denote by a, b, c their coordinates and let a', b', c' be the coordinates of their images A', B', C'. Then the relation

$$(A, B; C, M) = (A', B'; C', M')$$

can be rewritten as

$$\frac{c-a}{c-b} : \frac{x-a}{x-b} = \frac{c'-a'}{c'-b'} : \frac{x'-a'}{x'-b'}.$$

From this equation we can express x' in terms of x. The result looks like

$$x' = \frac{mx+n}{px+q}, \tag{42}$$

where m, n, p, q are certain constants depending on a, b, c, a', b', c'. Functions of this kind are called *fractional linear functions.*

Note that in this formula the expression $mq - np$ (determinant of the matrix $\left(\begin{smallmatrix} m & n \\ p & q \end{smallmatrix}\right)$) must be different from 0. Otherwise the pair (m, n) would be proportional to the pair (p, q), and the fraction will give one and the same value for all x.

Now suppose that m, n, p, q are four real numbers such that $mq - np \neq 0$. Is it true that the formula

$$f : x \mapsto \frac{mx + n}{px + q}$$

defines a one-to-one mapping of the real line $\mathbb{R}$ onto itself? The answer is negative: in fact, the point $x = -q/p$ (if $p \neq 0$) has no image under f.

Exercise 135. Indicate the real number that has no *inverse* image under this mapping.

We have already encountered these difficulties in one particular case (see page 140). The way out is to introduce one more point ∞ (infinity) and extend the action of the projective transformation to the set $\bar{\mathbb{R}} = \mathbb{R} \cup \infty$ using the rules:

- $\dfrac{a}{0} = \infty$ for any $a \neq 0$,
- $\dfrac{m \cdot \infty + n}{p \cdot \infty + q} = \begin{cases} \frac{m}{p}, & \text{if } p \neq 0, \\ \infty, & \text{if } p = 0. \end{cases}$

Fractional linear functions with $mq - np \neq 0$ define one-to-one transformations of the extended line $\overline{\mathbb{R}}$.

Exercise 136. Check that the set of all transformations given by formula (42) with $mq - np \neq 0$ forms a group.

This group is called *the group of projective transformations of the (extended) real line* and denoted by $\mathrm{PGL}(1, \mathbb{R})$.

Exercise 137. Check directly, using formula (42), that the cross ratio is an invariant of the projective group acting on the set of quadruples of points.

The argument that led us to formula (42) shows that any triple of distinct points can be taken into any other such triple by a suitable projective transformation. This means that the action of the projective group on the set of triples has no nontrivial invariants.

We have already several times considered the group generated by two projective transformations $x \mapsto 1/x$ and $x \mapsto 1 - x$ (see Exercise 74, Problem 43, etc.). This is not the only finite subgroup in the group of projective transformations.

Exercise 138. Prove that the two transformations $x \mapsto 1/x$ and $x \mapsto (x-1)/(x+1)$ generate a group of eight elements, isomorphic to D_4.

Exercise 139. Find all projective transformations of the line that have finite order.

That's all about projective transformations of the line. Now a few words about the plane.

The set of all projective transformations of the plane is a group denoted by $\mathrm{PGL}(2, \mathbb{R})$. It contains the set of all affine transformations $\mathrm{Aff}(2, \mathbb{R})$ as a subgroup.

Exercise 140. Is $\mathrm{Aff}(2, \mathbb{R})$ a normal subgroup of $\mathrm{PGL}(2, \mathbb{R})$?

Arguments similar to those that we used for projective transformations of the line imply the following theorems.

Theorem 14. *Projective transformations of the plane are those and only those transformations which are described by formulas*

$$(43) \qquad \begin{cases} x' &= \dfrac{a_1x + b_1y + c_1}{a_0x + b_0y + c_0}, \\ y' &= \dfrac{a_2x + b_2y + c_2}{a_0x + b_0y + c_0} \end{cases}$$

in a Cartesian (or affine) coordinate system.

Theorem 15. *The group* $\mathrm{PGL}(2, \mathbb{R})$ *acts transitively on the set consisting of all quadruples of points no three of which are collinear.*

The last theorem may turn out to be quite useful for solving some problems in elementary geometry. If a problem involves nothing but the incidence between points and lines, then we can make a projective transformation that takes any given quadrangle into another quadrangle, for which the solution might be easier. Remember, however, that projective transformations change not only the angles, distances and areas, but also ratios of segments on a line and ratios of areas of different figures. They only preserve straight lines and cross ratio.

Here is an example of such an application.

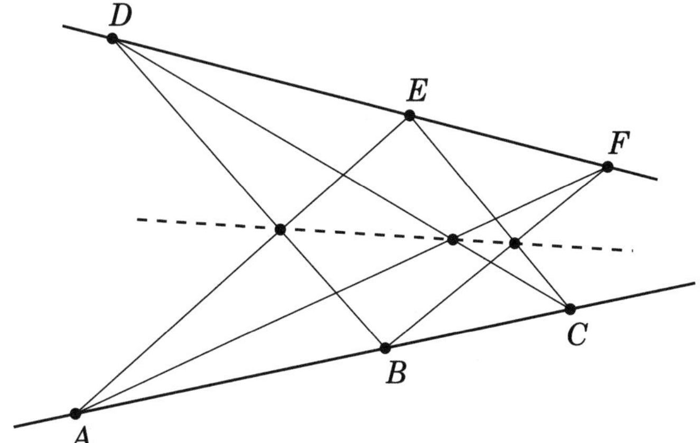

Figure 6. Pappus's theorem

Exercise 141. Prove the theorem of Pappus (see Figure 6): *if the three points* A, B, C *are collinear and the three points* D, E, F *are collinear, then the intersection points* $AE \cap BD$, $AF \cap CD$ *and* $BF \cap CE$ *are also collinear.*

3. Similitudes

Definition 37. A *similitude* is a plane transformation that changes all distances by one and the same positive factor.

Like affine transformations, similitudes constitute a class of plane transformations which is wider than the class of movements. It is clear from Definition 37 that the set of all similitudes is a transformation group.

The simplest type of similitudes, different from movements, is provided by *homotheties*.

Definition 38. A *homothety* H_A^k *with centre* A *and coefficient* $k \neq 0$ is the transformation that takes every point M into the point M' such that $\overrightarrow{AM'} = k \cdot \overrightarrow{AM}$ (see Figure 7).

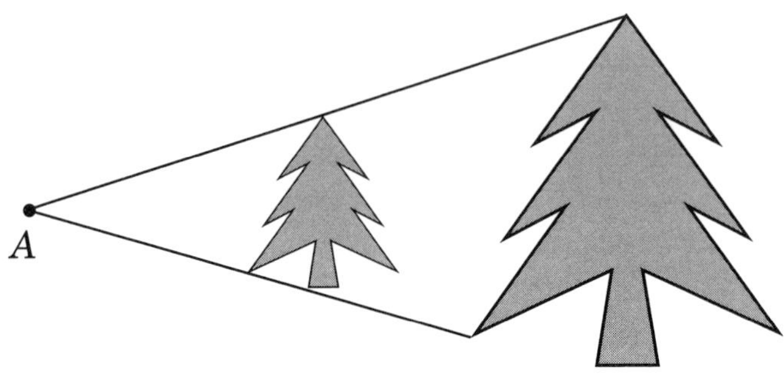

Figure 7. Homothety

The set of all homotheties of the plane does not constitute a group, but the set of all homotheties with a fixed centre does. Using a complex coordinate z, such transformations can be described by the formula $z \mapsto kz$, where k is a non-zero real number. This group is thus isomorphic to $\mathbb{R}^\star$, the multiplicative group of non-zero real numbers.

Exercise 142. Prove the isomorphism $\mathrm{GL}(2, \mathbb{R})/\mathbb{R}^\star \cong \mathrm{PGL}(1, \mathbb{R})$.

Below are some examples showing the use of homotheties in elementary geometry.

Problem 51. *In a given triangle ABC, inscribe a square in such a way that two of its vertices belong to one side of the triangle and the remaining two vertices lie on the other two sides.*

Solution. It is very easy to construct a square with three vertices satisfying the requirements of the problem (square $KLMN$ in Figure 8).

Any homothety centred at vertex A preserves these properties, and it remains to find the coefficient k so that H_A^k maps the point N into a point E belonging to the side BC. The whole construction is clear from the figure.

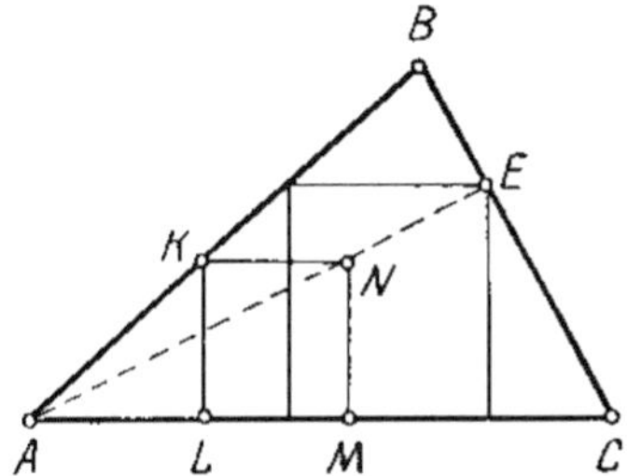

Figure 8. Inscribing a square into a triangle

Exercise 143. Into a given triangle, inscribe a triangle whose sides are parallel to the three given lines.

One more useful property of homotheties is that they preserve the direction of straight lines: the image of a line l is always a line parallel to l. We will use this fact in the following problem.

Problem 52. *Several circles are inscribed into one circular segment (Figure 9).*

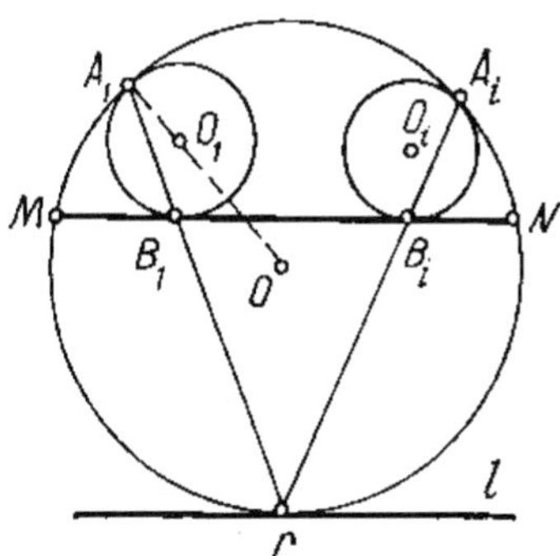

Figure 9. Circles inscribed into a circular segment

Let A_i *and* B_i *be the tangency points of the i-th inscribed circle with the arc and the chord, respectively. Prove that all lines* A_iB_i *pass through a common point.*

Solution. We will prove that every line A_iB_i passes through the point C which is the tangency point of the line l parallel to the chord MN and tangent to the big circle. Consider the homothety h_1 with centre A_1 and coefficient $k = OA_1 : O_1A_1$. It transforms the small circle S_1 (centred at O_1) into the big circle S. Therefore, the image of the line MN tangent to S_1 is the line l, tangent to S and parallel to MN. The point B_1, which is the common point of MN and S_1, goes under h_1 into the point C, common to l and S. The same argument can be repeated for each small circle S_i. This completes the proof.

Exercise 144. Given two concentric circles, construct a line which intersects them in the four consecutive points A, B, C, D so that the following relation holds for the lengths of the segments that are cut by the circles: $AB = 2BC = CD$.

Exercise 145. Given a triangle, prove that the three lines, each of which passes through the midpoint of a side parallel to the bisector of the opposite angle, meet in one point.

Exercise 146. Prove that for any triangle ABC there exists a circle that contains the midpoints of the sides, the feet of altitudes and the midpoints of the segments KA, KB, KC, where K is the intersection point of the altitudes. (This circle is called the *circle of 9 points*, or *Euler's circle*.)

The group of plane similitudes is not exhausted by the set of all homotheties.

Definition 39. A *spiral similarity* is defined as the composition of a homothety and a rotation with the same centre (see Figure 10).

In this book, we have already encountered such transformations when studying complex numbers: we saw that multiplication by a number a is equivalent to a homothety with coefficient $|a|$ and subsequent rotation through the angle $\arg a$ around the origin (see (4)). Let us consider some geometric applications of spiral similarities.

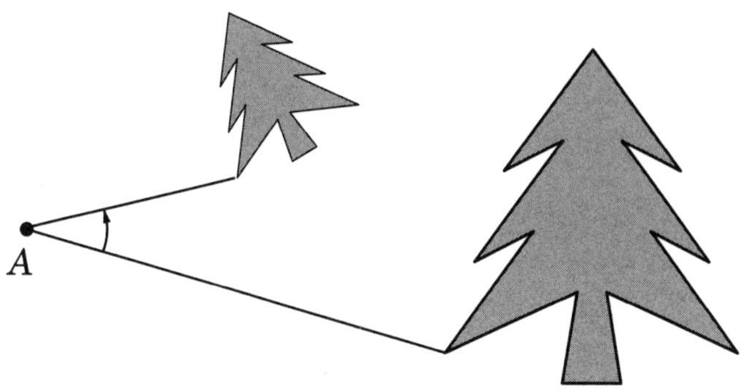

Figure 10. Spiral similarity

Problem 53. *Given an arbitrary triangle ABC, draw two triangles ABP and BQC, lying outside of $\triangle ABC$, having right angles at vertices P, Q and equal angles β at the vertex B (see Figure 11).*

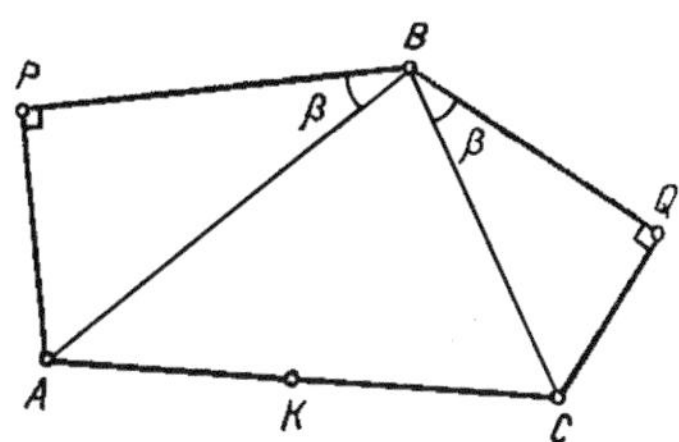

Figure 11. Right triangles built on the sides of triangle ABC

Find the angles of $\triangle PQK$, where K is the middle point of the side AC.

Solution. Consider two spiral similarities: $F_P = H_P^k \circ R_P^d$, $F_Q = H_Q^{1/k} \circ R_Q^d$, where $d = 90°$ and $k = PB : PA = QB : QC$. It is clear that $F_P(A) = B$ and $F_Q(B) = C$; hence $(F_Q \circ F_P)(A) = C$. When two spiral similarities are performed one after another their coefficients get multiplied, and their rotation angles are added (we will explain this a little later). Therefore, the composition

$F = F_Q \circ F_P$ must be a rotation through 180°. Since $F(A) = C$, the centre of rotation is the point K and thus $F(K) = K$. Let $F_P(K) = K_1$; then $F_Q(K_1) = K$. Both right triangles KPK_1 and QKK_1 have the same angle β at the vertex K_1; hence they are equal (Figure 12).

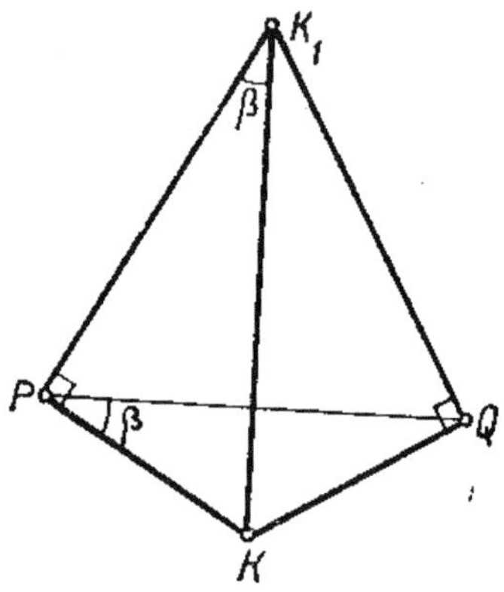

Figure 12. Product of two spiral similarities

It follows that $PQ \perp KK_1$ and $\angle KPQ = \angle KQP = \beta$.

In the previous argument, we have used the fact that the composition of two spiral similarities is a spiral similarity whose coefficient is the product of the two coefficients, while the angle of rotation is the sum of the two rotation angles. This fact is evident if the two transformations have a common centre. To prove it in full generality, we will use calculations with complex numbers, based on the following theorem.

Theorem 16. *A plane transformation is a similitude if and only if, in the complex coordinate z, it can be written as either*

$$z \mapsto pz + a \tag{44}$$

or

$$z \mapsto p\bar{z} + a, \tag{45}$$

where p and a are arbitrary complex numbers, $p \neq 0$. The two cases (44) and (45) correspond to proper (i.e., orientation-preserving) and improper (i.e., orientation-reversing) transformations.

Proof. Indeed, recall that we have already proved in (8) that proper movements of the plane correspond to the linear functions $w = pz + a$ with $|p| = 1$. Now, suppose that F is a proper similitude of the plane, i.e., a transformation that stretches all distances by a certain factor k and preserves the orientation. Let H be the homothety with coefficient k and centre 0. The composition $H^{-1} \circ F$ preserves the distances and the orientation; hence it is a proper movement and corresponds to a function $w = pz + a$ with $|p| = 1$. Then the transformation $F = H \circ (H^{-1} \circ F)$ can be written as $w = k(pz + a)$, which is an arbitrary linear function.

Conversely, given a complex function $pz + a$ with arbitrary coefficients, we can verify that it stretches the distances between points by the factor $k = |p|$:

$$|(pz_1 + a) - (pz_2 + a)| = |p| \cdot |z_1 - z_2|.$$

The case of improper transformations is reduced to the case of proper transformations by the simple observation that the function (45) is the composition of (44) and the standard reflection $z \mapsto \bar{z}$. □

In the terminology and notation of section 1, Theorem 16 means that the group of proper similitudes is $\mathrm{Aff}(1, \mathbb{C})$, where $\mathbb{C}$ is the group of complex numbers.

Using the description of similitudes in terms of complex numbers, we can easily prove two important facts, related to each other:

- Any transformation of similitude $pz+a$ which is not a translation (i.e., $p \neq 1$) has a unique fixed point.
- Any transformation of similitude $pz+a$ which is not a translation is a spiral similarity (in particular, a homothety).

Indeed, a fixed point is a number z_0 such that $pz_0 + a = z_0$. If $p \neq 1$, this equation has a unique solution $z_0 = a/(1-p)$. The formula

$$pz + a = p\left(z - \frac{a}{1-p}\right) + \frac{a}{1-p}$$

shows that this transformation is in fact a spiral similarity with centre $a/(1-p)$, stretching coefficient $|p|$ and rotation angle $\arg p$.

Now we can prove the fact used in Problem 53 above. The composition of two spiral similarities $w = pz + a$ and $u = qw + b$ corresponds to the function

$$u = q(pz + a) + b = pqz + (aq + b).$$

This is a spiral similarity with coefficient $|pq| = |p||q|$ and angle of rotation $\arg(pq) = \arg p + \arg q$.

Exercise 147. Two maps of the same country, drawn to different scales on transparent paper, are put on the table in such a way that one of the maps completely covers the other. Prove that one can pierce both maps with a pin in a point that corresponds to the same place on both maps.

Exercise 148. Given four points A, B, C, D in the plane, such that $\overrightarrow{AB} \neq \overrightarrow{CD}$, prove that there exists a point E for which the two triangles ABE and CDE are similar.

Exercise 149. Points M, N and P are centres of the squares constructed on the sides AB, BC, CA of an arbitrary triangle ABC outside of it. Prove that the segments NP and CM are perpendicular and have equal lengths.

4. Inversions

Problem 54. *A circle S touches two circles S_1 and S_2 at the points A and B. Prove that the line AB passes through the centre of similitude of the circles S_1 and S_2.*

Solution. Let K be the intersection point of the lines AB and O_1O_2 (see Figure 13). We want to prove that K is the centre of similitude of the circles S_1 and S_2. We will construct the required similitude in a rather indirect manner.

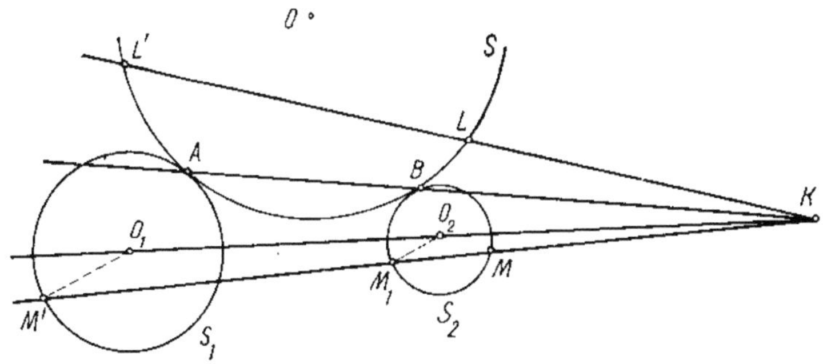

Figure 13. Three tangent circles

Let f be the transformation of the plane which, with every point M, associates the point M' belonging to the half-line KM at a distance from K that satisfies $KM \cdot KM' = KA \cdot KB = \text{const}$. Obviously, $f(A) = B$ and $f(B) = A$.

We claim that the circle S goes into itself under f. This follows from a well-known theorem of elementary geometry (if you don't know it, try to prove it yourself): *for a fixed circle S, a fixed point K and an arbitrary line l that passes through K and meets S at the two points L and L', the product of lengths of the two segments KL and KL' does not depend on the choice of the line l.*

Now take a point $M \in S_2$. Its image M' lies on the half-line KM and satisfies

$$KM' = \frac{KA \cdot KB}{KM}.$$

Let M_1 be the second intersection point of KM with S_2. By the theorem that we quoted, $KM \cdot KM_1 = C = \text{const}$. Therefore,

$$KM' = \frac{KA \cdot KB}{C} KM_1,$$

which means that M' is obtained from M_1 by a homothety centred at K! Therefore, the image of S_2 under

f is a circle, say S_2'. Since S_2 passes through B and is tangent to the circle S, the circle S_2' passes through the point A and is tangent in that point to S; thus $S_2' = S_1$.

We have proved that the two circles S_1 and S_2 can be transformed into one another by a homothety with centre K.

The transformation f that we used in the previous problem is an example of inversion.

Definition 40. The *inversion with respect to the circle T with centre O and radius* r is the transformation that maps every point M into the point M' that belongs to the half-line OM and satisfies the equality $OM \cdot OM' = r^2$.

The inversion map takes the inside of the circle outside and the outside inside. It preserves the circle itself. A well-known joke of H. Pétard ("A contribution to the mathematical theory of big game hunting") suggests the following method to catch a lion. The hunter gets into a cage and waits. When the lion appears, he performs an inversion. Now the lion is inside the cage.

Inversion is an *almost* one-to-one transformation of the plane: it is defined and one-to-one everywhere except at the centre of the circle O. When the point M moves towards O, its image M' moves infinitely far from O. This is why it is natural to add the point ∞ ("infinity") to the plane, similarly to what we did for projective transformations of the line on page 173, and consider inversion as a one-to-one transformation of the extended plane.

Suppose that our plane is the plane of complex numbers. We know that every complex number z and its conjugate $\bar{z}$ satisfy the relation $z\bar{z} = |z|^2$. Therefore, the algebraic formula for an inversion of radius r with centre 0 is $z \mapsto r^2/\bar{z}$.

Exercise 150. What transformation group is generated by the set of all inversions with a fixed centre 0?

During the discussion of Problem 54, we found that an arbitrary circle that does not pass through the centre of the inversion f maps under f into a certain circle.

Figure 14. Use of inversion to catch a lion (after H. Pétard)

Exercise 151. What is the image, under an inversion, of a circle that passes through the centre of inversion?

Exercise 152. What is the image, under an inversion, of a straight line?

All these facts, put together, mean that *the inversion preserves the set of all lines and circles.* Viewing a straight line as a circle passing through infinity, we can say that inversions are *circular transformations*, i.e. transformations that preserve the class of all (generalized) circles. A little later we will see that the set of circular transformations is not exhausted by inversions.

Now we give some more applications of inversions in elementary geometry.

Problem 55. *Each of four circles touches two of its neighbours (Figure 15a). Prove that the points of contact lie on one circle.*

Solution. Let us apply an inversion with centre A (one of the contact points) and an arbitrary radius. We will

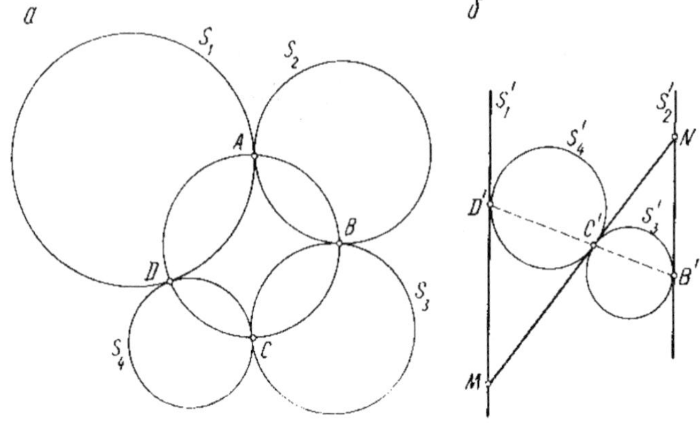

Figure 15. Four circles touching each other

see that the problem becomes simpler after this transformation.

Denote by S_i' the image of S_i. It follows from our previous considerations that S_1' and S_2' are straight lines, and S_3' and S_4' are circles. The relations of tangency between S_i' are the same as between S_i, i.e., S_1' is tangent to S_2', S_2' to S_3', S_3' to S_4', and S_4' to S_1'. Note also that the lines S_1' and S_2' must be parallel, because S_1 and S_2 have only one common point A, which goes to infinity under the inversion. We arrive at the configuration shown in Figure 15b. The problem is to prove that the three points of contact B', C', D' belong to one straight line — this will imply that the inverse images of these points B, C, D belong to a circle that passes through the centre of the inversion A.

To prove that $B'C'D'$ is a straight line, let us draw the common tangent of the circles S_3' and S_4' until the intersection with the lines S_1' and S_2' in the points M and N. Consider the two triangles $MC'D'$ and $NC'B'$. They are equilateral and have equal angles $\angle M = \angle N$. Therefore their angles at the vertex C' are also equal to each other. Hence $B'C'D'$ is a straight line. The proof is complete.

In problems 54 and 55 we used the evident property that if two lines are tangent to each other, then their images under the inversion are also tangent. The next exercise is a generalization of this fact.

Exercise 153. Define the angle between the two circles at a point of intersection to be the angle made by their tangent lines drawn through that point. Prove that inversion with respect to any circle preserves angles between circles.

5. Circular transformations

The set of all inversions in the plane is not a transformation group. In this section, we shall study the group *generated* by all inversions. This group is called *the group of circular transformations.* It consists of two halves: the subgroup of orientation-preserving transformations, which coincides with the complex projective group $\mathrm{PGL}(1, \mathbb{C})$ (see page 173), and a coset consisting of orientation-reversing transformations.

We start with an illustrative problem.

Problem 56. *Fix a circle C with centre O. Let A be the midpoint of its radius OB. Suppose that we are allowed to perform two transformations: inversion with respect to the circle C and half turn around point A. What is the maximal number of different points that can be obtained from a given point by successive applications of these transformations?*

Solution. Let us write both transformations as functions of a complex variable, assuming that the point O has complex coordinate 0 and point B complex coordinate 1.

The point A corresponds to the number $1/2$, and the symmetry at this point is described by the function

$$f_1(z) = 1 - z.$$

To find the formula for the second allowed transformation, note that points z and w that correspond to each other under the inversion satisfy the two relations

$|z||w| = 1$ and $\arg z = \arg w$. It follows that $w = 1/\bar{z}$, and thus

$$f_2(z) = 1/\bar{z}.$$

Each of the two allowed transformations is involutive, and so the only way to obtain different compositions of the two functions is to apply them by turns. Starting from z and applying first f_1, then f_2, then again f_1, etc., we obtain the following list: z, $1-z$, $1/(1-\bar{z})$, $\bar{z}/(\bar{z}-1)$, $1-1/z$, $1/z$, $\bar{z}$, $1-\bar{z}$, $1/(1-z)$, $z/(z-1)$, $1-1/\bar{z}$, $1/\bar{z}$. After this we obtain z once again, and the sequence begins looping. Therefore, the inversion and the central symmetry generate a group G of 12 elements, and its orbit cannot contain more than 12 points. An example where the orbit contains exactly 12 points is given below.

Exercise 154. Find all the possibilities for the number of points in the orbits of the group G. Draw pictures of different types of orbits.

Now let us find the *fundamental domain* of the group G, i.e., the part of the plane whose images under the group action cover the plane without overlapping. The image of the circle C under the transformation f_1 is the circle C' (see Figure 16).

The image of C' under f_2 is the line MM'. One more important line is the straight line OB which separates the two regions corresponding to each other under the complex conjugation $z \mapsto \bar{z}$ (a mapping belonging to our group). These lines divide the plane into 12 domains that go into one another under the group action. Each of these domains has the property that it does not contain interior points equivalent under G.

Any of the 12 domains (e.g., domain 1) can be taken as the fundamental domain of the group action under study. The reader is invited to check what are the images of domain 1 under different transformations of the group.

The union of the four lines drawn in Figure 16 is the set of all points on the plane left invariant by some nontrivial element of the group. The orbit of any interior point consists of exactly 12 points.

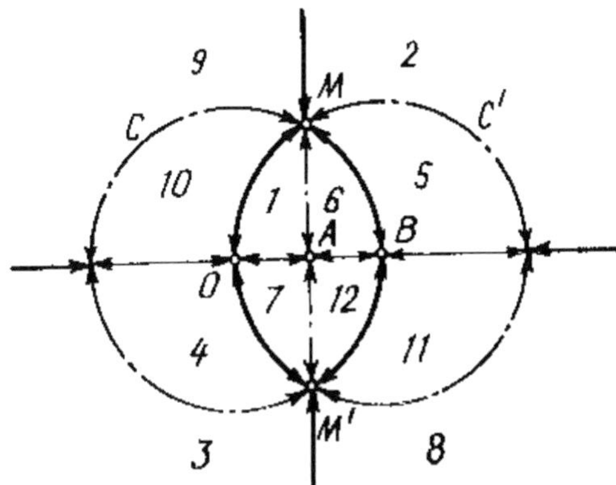

Figure 16. Group of complex functions generated by $z \mapsto 1 - z$ and $z \mapsto 1/\bar{z}$.

The orbits of the points of the lines different from their intersection points have cardinality 6. The intersection points split into three orbits: two of length 3 ($1/2, 2, -1$ and $0, 1, \infty$) and one of length 2 (points M and M', corresponding to the numbers $1/2 \pm i\sqrt{3}/2$).

The four lines shown in Figure 16 are divided into 18 segments by the intersection points. These segments go into one another by the group action and split into 3 orbits of length 6, shown in the figure as normal, bold and dotted lines.

Now we shall study the geometric meaning of plane transformations described by fractional linear functions with arbitrary complex coefficients, i.e., elements of the group $\mathrm{PGL}(1, \mathbb{C})$, as well as similar functions with z replaced by the conjugate variable $\bar{z}$.

Theorem 17. *Let a, b, c, d be any complex numbers such that $ad - bc \neq 0$. Then:*

(1) *The transformation defined by the function*

$$w = \frac{a\bar{z} + b}{c\bar{z} + d}$$

(an improper fractional linear transformation) is a composition of an inversion and a spiral similarity.

(2) *The transformation defined by the function*

$$w = \frac{az + b}{cz + d}$$

(a proper fractional linear transformation) is a composition of an inversion, a spiral similarity and a reflection.

Proof. If $c = 0$, then the second formula gives a linear function $a_1 z + b_1$, which, as we know, corresponds to a similitude transformation. The first formula gives $a_1 \bar{z} + b_1$, a linear function in $\bar{z}$, which is the composition of the reflection $z \mapsto \bar{z}$ and a similitude.

Suppose now that $c \neq 0$. Then the fraction $\dfrac{a\bar{z} + b}{c\bar{z} + d}$ can be written as

$$\frac{a\bar{z} + b}{c\bar{z} + d} = p\left(\frac{1}{\bar{z} - \bar{z}_0} + z_0\right) + r,$$

where $z_0 = -\overline{d}/\overline{c}$, $p = (bc - ad)/c^2$ and $r = a/c - pz_0$. The expression inside the parentheses is the conjugation of the standard inversion (with centre 0 and radius 1) by the translation $z \mapsto z + z_0$; therefore, it represents the inversion of radius 1 centred at the point z_0. To the result of the inversion, the similitude transformation $z \mapsto pz + r$ is applied, and we get the required composition.

The proper fraction $(az + b)/(cz + d)$ is reduced to the improper one by the change $z \mapsto \bar{z}$, and we obtain the second part of the theorem. It is funny that in this case improper transformations, those that change orientation, are easier to handle than the proper transformations. This observation is accounted for by the importance of inversions in this context — and inversions are improper transformations. □

Exercise 155. Check that all (proper and improper) fractional linear transformations form a group, and the set of proper transformations is a normal subgroup of it. Find the quotient group.

Theorem 17 implies that both classes of fractional linear transformations are *circular*, i.e., they preserve the set of generalized circles (circles and straight lines) in the plane. It is also true that any circular transformation is described by either a proper or an improper fractional linear function. This is why the group of all such transformations is called *the circular group*. Another noteworthy property of these transformations is that they are *conformal*, i.e., they preserve angles between curves. However, the class of all conformal mappings is much wider than that of circular transformations — for example, it includes complex functions $P(z)/Q(z)$, where P and Q are arbitrary polynomials.

Exercise 156. Prove that all transformations of the complex plane given by formulas

$$w = \frac{az+b}{cz+d}, \quad a, b, c, d \in \mathbb{R}, \quad ad - bc > 0, \tag{46}$$

and

$$w = \frac{a\bar{z}+b}{c\bar{z}+d}, \quad a, b, c, d \in \mathbb{R}, \quad ad - bc < 0, \tag{47}$$

form a group.

6. Hyperbolic geometry

Let us check that transformations (46) and (47) map the upper half-plane $y > 0$ into itself. If $z = x + iy$, $w = u + iv$, then a simple calculation shows that the complex formula (46) is equivalent to the pair of real formulas

$$\begin{aligned} u &= \frac{(ax+b)(cx+d)+acy^2}{(cx+d)^2+y^2}, \\ v &= \frac{(ad-bc)y}{(cx+d)^2+y^2}, \end{aligned}$$

and we see that v has the same sign as y. The formula (47) is considered in a similar way.

Let L be the group of all transformations (46) and (47) acting on the upper half-plane $H = \{(x, y)|y > 0\}$. The half-plane H is called the *hyperbolic plane*, or the *Lobachevsky plane*, and the group L is called the group of *hyperbolic movements* of H. This terminology has the following meaning.

As we know, the transformations in L take any circle into a circle (or a line, which we view as a particular case of the circle). In the plane H, there is a distinguished set of circles which is preserved by the group L. These are the (half)-circles and (half)-lines perpendicular to the line Ox (see Figure 17). We will call these circles the *L-lines*, because through any two points of H there passes one and only one L-line — a property owned also by the set of all usual lines in the usual plane.

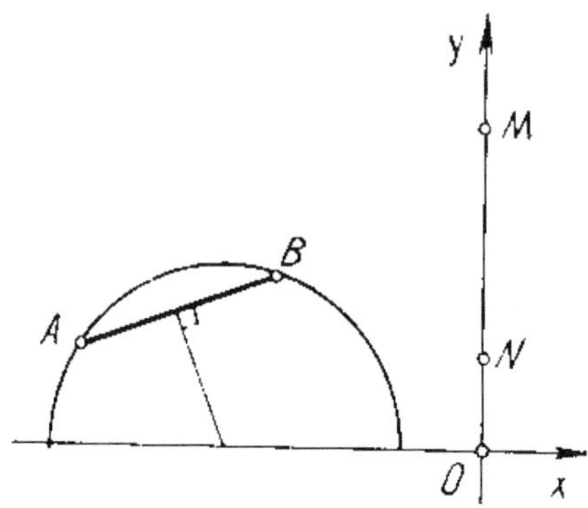

Figure 17. L-lines in hyperbolic plane

The group L of hyperbolic movements, or L-*movements*, has properties similar to those of the group of plane movements acting on the usual plane. In particular, any point can be taken into any other by a hyperbolic movement, but the action of L on the set of L-segments (arcs of L-lines) is not transitive. The main geometric difference between the hyperbolic and the usual planes appears when we think about parallel lines.

In ordinary Euclidean geometry, two lines are called parallel if they do not have common points, and the main property of parallel lines is that for any line a and any point A outside of a there is exactly one line passing through A and parallel to a. Now look at Figure 18, which shows an L-line l and an L-point A. Among the four lines drawn through A, there is one (l) that intersects the line a, and there are three (k, n, m) that have no common points with a (we recall that the points of the boundary horizontal line do not belong to H). We thus see that in Lobachevsky geometry one can draw many lines passing through the given point and not intersecting the given line.

Let us do a computational exercise in Lobachevsky geometry. The angle between two L-lines is by definition measured as the usual

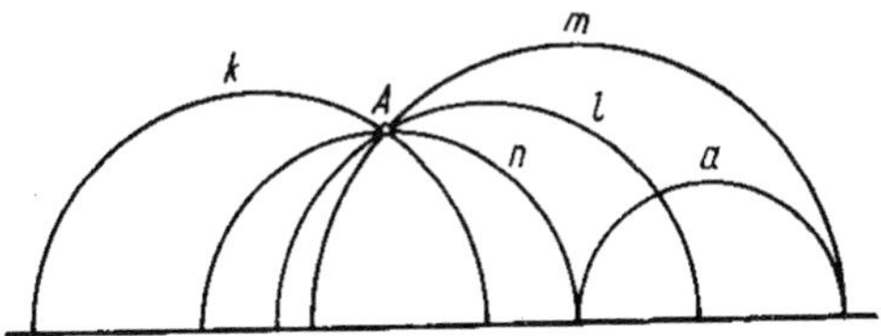

Figure 18. Mutual position of two L-lines

Euclidean angle between the tangent lines (note that with this definition of the angle we have the property that L-movements preserve L-angles).

Problem 57. *Find the sum of angles of the Lobachevsky triangle with vertices* $A(0,7)$, $B(4,3)$, $K(0,5)$.

Solution. The given triangle ABK is the hatched region in Figure 19.

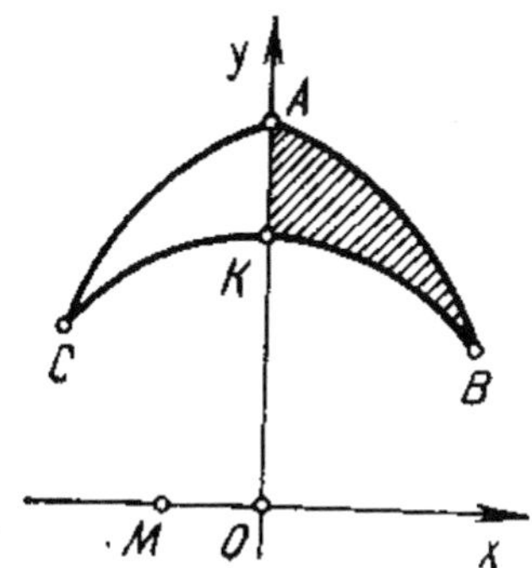

Figure 19. A triangle in the hyperbolic plane

The side AK lies on the axis Oy. The side KB is an arc of a circle with centre at O. The side AB is an arc of the circle with centre at the point M.

Exercise 157. Find the coordinates of the point M.

The angle K of our triangle is a right angle, because it is formed by a circle and its radius:

$$\angle K = 90^\circ.$$

The angle B between the two circles is equal to the angle between the tangents and hence to the angle between the radii:

$$\angle B = \angle OBM.$$

Similarly,

$$\angle A = \angle OMA.$$

If you have found the coordinates of M, then you can find that

$$\begin{aligned} \tan \angle OMA &= \frac{7}{3}, \\ \tan \angle OBM &= \tan(\angle BOB_1 - \angle BMB_1) = \frac{9}{37}, \\ \tan(\angle A + \angle B) &= \frac{\frac{7}{3} + \frac{9}{37}}{1 - \frac{7}{3} \cdot \frac{9}{37}} = \frac{143}{24}. \end{aligned}$$

Since the tangent is positive, we infer that $\angle A + \angle B < 90^\circ$. Therefore, the sum of the three angles of the triangle ABK is less than 180°.

It is interesting to note that the bigger the Lobachevsky triangle (in a certain sense), the smaller its sum of angles. For example, you can check that the isosceles triangle ABC, which is twice the triangle ABK, has a smaller sum of angles.

Finally, we will give one example of a crystallographic group in the hyperbolic plane — the so-called *modular group* U that consists of all proper fractional linear transformations with integer coefficients:

$$U = \left\{ \frac{az + b}{cz + d} \,\middle|\, a, b, c, d \in \mathbb{Z} \right\}.$$

This group is generated by the two elements

$$S : z \mapsto -1/z \quad \text{and} \quad T : z \mapsto 1 + z.$$

Exercise 158. Check the relations $S^2 = (ST)^3 = \mathrm{id}$.

Figure 20 shows the fundamental domain of the group U

$$\Phi = \{z = x + yi \mid |z| \geq 1,\ |x| \leq \tfrac{1}{2}\}$$

and its images under T, S, T^{-1}, TS, ST, etc.

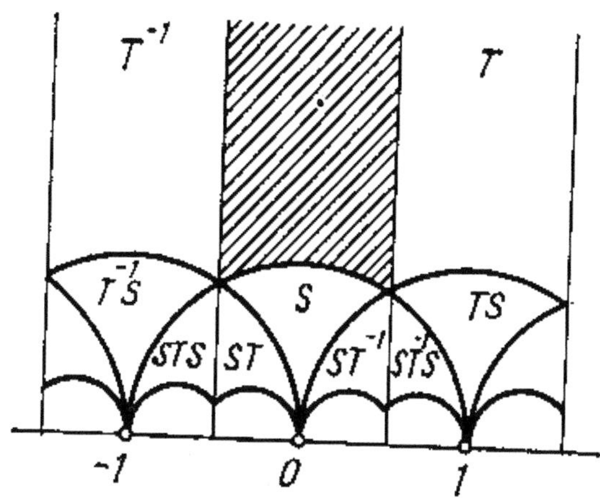

Figure 20. Fundamental domain of the modular group

The domain Φ is in fact an L-triangle, and it has a finite area (we have not defined area in Lobachevsky geometry, so you cannot check that!). Copies of Φ cover all the upper half-plane without overlapping. Thus U is really a crystallographic group. The reader is invited to draw a motif and repeat it throughout the Lobachevsky plane using the action of the group U, and thus obtain a *hyperbolic ornament.*

Chapter 7

Symmetries of Differential Equations

In this chapter we will apply the machinery of transformation groups to the solution of differential equations. We assume that the reader is acquainted with the notions of derivative and of definite and indefinite integral.

1. Ordinary differential equations

In this book we will only study the simplest class of differential equations: *ordinary differential equations of first order resolved with respect to the derivative.*

Definition 41. A *differential equation* is an equation of the form

$$y' = f(x, y), \tag{48}$$

where y is a variable depending on x, the prime means the derivative with respect to x, and $f(x, y)$ is a given function of two variables, x and y, which is supposed to be "good enough" (continuous and differentiable).

Definition 42. A *solution to equation* (48) is a function $y = \phi(x)$ which, upon substitution into the equation, makes it a true identity, so that

$$\phi'(x) = f(x, \phi(x))$$

for any value of x.

Since we are interested in the way y depends on x, we call x the *independent* and y the *dependent* variable. Equation (48) can also be written as $dy/dx = f(x, y)$, where dy and dx are *differentials*, i.e. infinitesimal ("infinitely small") increments of y and x whose ratio is by definition equal to the derivative y'. [1]

Here is an example of a differential equation:

$$y' = y - x. \tag{49}$$

As you can check by a direct substitution, either of the functions $y = x + 1$ and $y = e^x + x + 1$ is a solution of this equation.

The main theorem of the theory of ordinary differential equations implies that every differential equation has a one-parameter family of solutions that can be described by a formula $y = \psi(x, c)$ containing a constant c whose value may be arbitrary. Such a function $\psi(x, c)$ is called the *general solution* of the given equation. For example, equation (49) has the general solution $y = ce^x + x + 1$ which gives the two particular solutions quoted above, when $c = 0$ and $c = 1$.

Note that the family of solutions of a differential equation $\psi(x, c)$ cannot be an arbitrary one-parameter family of functions.

Exercise 159. Is there a (first order) differential equation that has the following pair of particular solutions: (a) $y = 0$ and $y = 1$ (constant functions)? (b) $y = 1$ and $y = x$?

The function $f(x, y)$ in the right-hand side of the differential equation (48) can be free of x, of y, or of both. For example, we can consider the following equations:

$$y' = 2, \tag{50}$$

$$y' = \cos x, \tag{51}$$

$$y' = y^2. \tag{52}$$

Exercise 160. Find the general solution of (50) and (51). Try to guess a particular solution of (52).

[1] From the modern viewpoint, the notion of differential is formalized using *differential forms*, but, as we cannot touch upon that in this book, we are going to treat the differentials in the above intuitive sense, following the mathematicians of the seventeenth and eighteenth centuries.

Equations (50) and (51) belong to the class of equations whose right-hand sides depend only on x:

$$y' = f(x). \tag{53}$$

The reader knows that the general solution to such an equation is obtained by indefinite integration:

$$y = \int f(x)dx, \tag{54}$$

where the right-hand side is defined "up to an additive constant". More exactly, if $F(x)$ is a certain *primitive* of $f(x)$, i.e., a function such that $F'(x) = f(x)$, then the general solution to (53) can be written as

$$y = F(x) + C. \tag{55}$$

This formula, for arbitrary values of the constant C, gives all the solutions of (53). The graphs of all functions (55) do not intersect and fill all the plane (x, y). For example, for (51) we obtain the picture shown in Figure 1a.

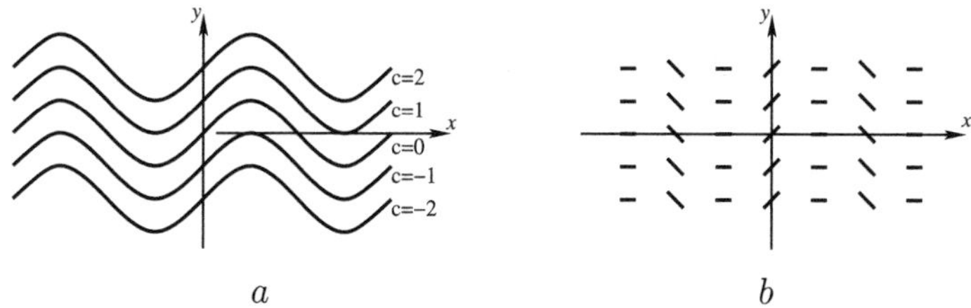

Figure 1. Graphs of solutions and field of directions of a differential equation

Not only the set of solutions, but the differential equation itself can be represented as a geometric object. The equality $y' = f(x, y)$ means that the slope (more exactly, the tangent of the slope angle) of the graph of the unknown solution at the point (x, y) should be equal to the known number $f(x, y)$. Therefore, at every point of the plane (x, y) we know the direction in which the integral curve (the graph of a solution) should pass. We arrive at the following conclusion: *the geometric object associated with the differential equation* (48) *is a field of directions in the plane.* A field of directions is fixed whenever, for every point of the plane, one defines a line passing through that point.

Geometrically, the problem of integrating a differential equation is formulated as follows: *given a field of directions in the plane, find all the curves that are everywhere tangent to the given field.* Such curves are referred to as *integral curves* of the field of directions. Figures 1 and 2 show the direction fields and the families of integral curves corresponding to equations (51) and (49), respectively.

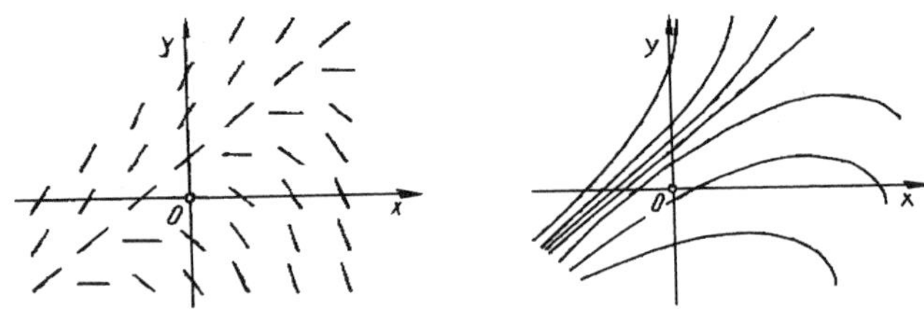

Figure 2. Field of directions and solutions of another differential equation

Exercise 161. Draw the direction fields and the families of integral curves for equations (50) and (52).

Exercise 162. What are the integral curves of the field of directions shown in Figure 3? Does this field correspond to any differential equation?

Using only indefinite integration, one can solve not only equations of class (53) (independent of y), but also equations of the form

$$y' = f(x)g(y), \tag{56}$$

This can be done by the following classical trick. Write y' as the ratio of two differentials dy/dx and then rewrite (56) as

$$\frac{dy}{g(y)} = f(x)\,dx. \tag{57}$$

As you see, the variables are separated: on the left, we have only y, on the right, only x. This is why equations of type (56) are called

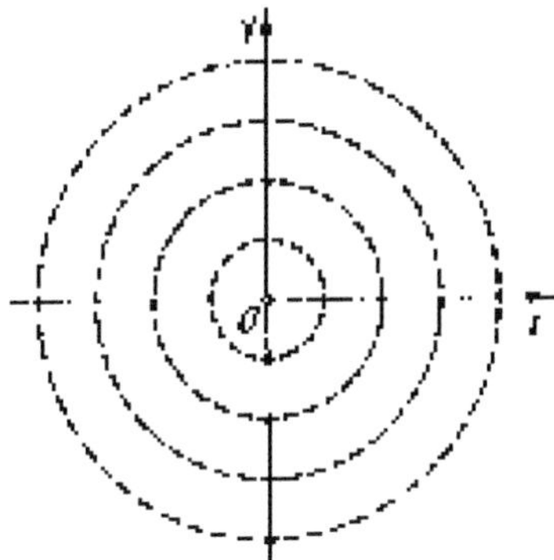

Figure 3. A field of directions in the plane

equations with separating variables. Integrating both sides of (57), we obtain

$$\int \frac{dy}{g(y)} = \int f(x)\, dx \tag{58}$$

(it is understood that one side of this equality contains an arbitrary additive constant C). This is an implicit formula for the general solution of (41). If y is expressed in terms of x, we will get an explicit general solution.

Problem 58. *Find the general solution of the differential equation*

$$y' = (2x + 1)/(3y^2). \tag{59}$$

Solution. Rewrite the equation in terms of differentials: $3y^2\, dy = (2x+1)\, dx$. Finding the indefinite integral gives $y^3 = x^2 + x + C$, whence $y = \sqrt[3]{x^2 + x + C}$. This is the general solution of Equation 59.

Special cases of equations with separating variables consist of equations whose right-hand sides depend on only one variable, x or y.

Exercise 163. Find the general solution of equation (52).

2. Change of variables

We now discuss of the relation between the differential equations and the main theme of the book — transformations of the plane.

It turns out that in various methods of finding solutions of differential equations a crucial role is played by *changes of variables*. If we pass from variables x, y to new variables u, v according to some formulas

$$\left\{ \begin{array}{lcl} u &=& \varphi(x,y), \\ v &=& \psi(x,y), \end{array} \right. \tag{60}$$

then the equation $y' = f(x,y)$ transforms to another equation

$$v' = g(u,v) \tag{61}$$

where the prime means the derivative with respect to u, not x. If it turns out that in this equation variables separate, then we can solve it and then, using (60), return to the initial variables x, y and obtain the solution of the initial equation.

What we have described is a very simple, but very effective method of integration: make a change of variables that leads to an equation with separating variables.

Problem 59. *Solve the equation $y' = y - x$ (49) by reducing it to an equation with separating variables.*

Solution. Let us make the following change of variables:

$$\left\{ \begin{array}{lcl} u &=& x, \\ v &=& y - x - 1. \end{array} \right.$$

Since $u = x$, the derivative with respect to u is the same thing as the derivative with respect to x; therefore no confusion arises if we denote both of them by a prime. Next we have $y' = v' + 1$ and, substituting this into the given equation, we obtain the equation $v' = v$. This is an equation with separating variables, it has the general solution $v = Ce^u$. Coming back to the variables x, y, we get the general solution of the initial equation in the form $y = Ce^x + x + 1$.

Exercise 164. Find the change of variables that transforms the equation $y' = y^2+2xy+x^2-1$ to an equation with separating variables.

The formulas of the change of variables (60) have a double geometric meaning.

First, regarding x, y as Cartesian coordinates of a point in the plane, we can view u, v as the coordinates of the same point in another curvilinear coordinate system. For example, the formulas

$$\begin{cases} u &= \sqrt{x^2+y^2}, \\ v &= \arctan\dfrac{y}{x} \end{cases}$$

or the equivalent formulas

$$\begin{cases} x &= u\cos v, \\ y &= u\sin v \end{cases}$$

introduce the system of polar coordinates (u, v).

Second, we can think that we deal with a transformation of the plane, which takes a point with coordinates (x, y) into the point with coordinates (u, v) where $u = \varphi(x, y)$, $v = \psi(u, v)$. All the coordinates are in this case calculated in one and the same coordinate system. To visualize the transformation in this case, it is useful to draw the images of the coordinate lines $x = \text{const}$ and $y = \text{const}$.

3. The Bernoulli equation

Historically, the first person who successfully applied transformations of variables to differential equations was probably Johann Bernoulli, who solved the equation (now bearing his name)

$$y' = Ay + By^n, \tag{62}$$

where A and B are given functions of x. He managed to reduce this equation to a simpler (*linear*) equation

$$y' = Py + Q, \tag{63}$$

where P and Q are again functions of x.

Let us first explain how the linear equation is solved. Write the unknown function y as a product $y = uv$, where u and v are unknown functions of x. Substituting this into (63), we get

$$u'v + uv' = Puv + Q.$$

This equation is satisfied, if the following two relations hold: $u' = Pu$, $v' = Q/u$. The first is an equation with separating variables, from which we can find the function $u(x)$. Feeding it into the second one, we can find $v(x)$ by simple integration. We thus get the solution of the initial equation $y = u(x)v(x)$.

Exercise 165. Find the general solution of the equation

$$y' = 2\frac{y}{x} - x^3 + x.$$

Problem 60. *Find a transformation that reduces Bernoulli's equation* (62) *to the linear equation* (63).

Solution. Equations (62) and (63) differ only in the exponents of the dependent variable y. Therefore, it is natural to try a transformation of the form $y = v^k$ (preserving the independent variable x). Let us substitute this expression into the equation and see what happens:

$$kv^{k-1}v' = Av^k + Bv^{kn},$$

or

$$v' = \frac{A}{k}v + \frac{B}{k}v^{kn-k+1}.$$

If $k = 1/(1-n)$, the second exponent $kn - k + 1$ becomes 0 and we arrive at a linear equation! Therefore, the required transformation is $y = v^{1/(1-n)}$.

Exercise 166. Find the general solution of the equation

$$y' = \frac{xy^2 + 1}{2y}.$$

A reader who has solved this exercise (or looked at the answer in the back of the book) might be perplexed by the fact that the solution is not given by a conventional formula, as a closed expression in elementary functions. We must therefore say a few words about *integration in closed form*. The function

$$\int e^{-x^2/2}dx,$$

although it is not an elementary function (it cannot be written as a combination of polynomials, trigonometric functions, logarithms and exponents), is in fact almost as good as any elementary function. Its numeric values can be found by computer to any degree of precision and its properties are well known, because this function is widely used in probability theory and statistics. The same refers also to the integral of any elementary function. This lays the groundwork for the following important notion.

A differential equation is said to be *integrable in closed form*, if its general solution can be written by a formula involving elementary functions, indefinite integrals and inverse functions.

The simplest example of an equation which is not integrable in closed form is the equation

$$y' = y^2 + x$$

(this fact was proved by J. Liouville in 1841). This equation is a particular case of the so-called *Riccati equation*

$$y' = a(y^2 + x^n). \tag{64}$$

Exercise 167. Find the general solution of the Riccati equation (64) for $n = 0$.

In the year 1742, D. Bernoulli and J. Riccati discovered a discrete series of values of the parameter n for which the equation (64) can be integrated in closed form. This was done by a very elegant trick, actually by means of a cyclic group of transformations. The idea was to find a change of variables which takes (64) into an equation of the same form, but with a different value of the exponent n, and then try to reduce the equation to the case $n = 0$ (which is integrable, as you know from Exercise 167).

Let us first make the change of the dependent variable according to the formula

$$y = \frac{1}{x^2 v} - \frac{1}{ax}.$$

Feeding this into (64), after some simplifications we obtain an equation for v:

$$v' = a\left(-\frac{1}{x^2} - x^{n+2}v^2\right), \tag{65}$$

where the prime, as before, means the derivative with respect to x.

This is not yet a Riccati equation, but we will get one if we change the independent variable according to the rule[2] $u = x^{n+3}$ (or $x = u^{1/(n+3)}$). Indeed, by the chain rule we have

$$v' = \frac{dv}{dx} = \frac{dv}{du}\frac{du}{dx} = \frac{dv}{du}(n+3)x^{n+2} = (n+3)u^{\frac{n+2}{n+3}}\frac{dv}{du}.$$

A simple calculation shows that after this change equation (65) becomes

$$\frac{dv}{du} = -\frac{a}{n+3}\left(v^2 + u^{-\frac{n+4}{n+3}}\right).$$

This is again a Riccati equation, but with the exponent n changed to $-(n+4)/(n+3)$.

If, for example, we had an equation with $n = -4$, after this transformation we would obtain the equation with $n = 0$ — which is integrable. Therefore, Riccati's equation with $n = -4$ is integrable, too.

Exercise 168. Find the general solution of the equation $y' = y^2 + x^{-4}$.

Exercise 169. Find one more value of n for which the Riccati equation is integrable in closed form.

Now let us make our observations into a general theory. We know that, if the Riccati equation is integrable for a certain exponent m, it is also integrable for the exponent n such that $-(n+4)/(n+3) = m$, i.e. $n = -(3m+4)/(m+1)$.

Consider the fractional linear function

$$q(m) = -\frac{3m+4}{m+1}.$$

By the previous argument, if the Riccati equation (64) is integrable for some exponent m, then it is also integrable for the value $q(m)$. Repeating the transformation, we deduce that it is also integrable for the exponents $q(q(m))$, $q(q(q(m)))$ and in general for any $q^k(m)$, where q^k means the k-th power of the transformation q.

Exercise 170. Find an explicit formula for $q^k(m)$.

---

[2]We only quote the transformations invented by Bernoulli and Riccati. Nobody knows *how* they found them!

Note that the inverse transformation q^{-1} has the same property: it takes an "integrable" exponent into an "integrable" exponent. We obtain an infinite cyclic group generated by the fractional linear transformation q. This group acts on the set of all real numbers (exponents of the Riccati equation). The property of the equation to be integrable in closed form is an invariant of this action. Therefore, each orbit either consists entirely of exponents for which the equation is integrable, or contains only such exponents for which the equation is not integrable.

In particular, the orbit of the number 0 furnishes an infinite series of Riccati equations that are integrable in closed form. Their exponents are

$$q^k(0) = \frac{4k}{1-2k},$$

where $k \in \mathbb{Z}$ is an arbitrary integer. Note that $q^k(0)$ tends to the value -2, when k goes to infinity.

Exercise 171. Prove that the Riccati equation (64) is also integrable for the exponent $n = -2$.

We must, however, warn the reader that, starting from the "integrable" value $n = -2$ and using the transformation q, it is impossible to find any new integrable cases, because the number -2 is a fixed point of q and its orbit consists of only one point.

We have thus found two integrable orbits in the set of Riccati exponents. J. Liouville proved that for all the remaining values Riccati's equation cannot be solved in closed form.

4. Point transformations

So far, we have only encountered changes of variables of the form

$$\begin{cases} x &= \varphi(u), \\ y &= \psi(u,v), \end{cases} \tag{66}$$

i.e., where the independent variable x is expressed in terms of the new independent variable only. In this case it is easy to express the derivative dy/dx in terms of u, v and dv/du using the chain rule $dy/dx = dy/du \cdot du/dx$.

One can, however, use arbitrary transformations of the independent and dependent variables $x = \varphi(u, v)$, $y = \psi(u, v)$ (*point transformations*). To derive the transformation formula for the derivative dy/dx in this case, we will need the notion of *partial derivatives.*

Definition 43. Let $z = h(x, y)$ be a function of two variables. If the value of y is fixed, $y = y_0$, we obtain a function of one variable $z = h(x, y_0)$. The derivative of this function at the point x_0 is called *the partial derivative of the function $h(x, y)$ with respect to the variable x at the point* (x_0, y_0). The partial derivative is denoted by $\frac{\partial h}{\partial x}(x_0, y_0)$. Symbolically,

$$\frac{\partial h}{\partial x}(x_0, y_0) = \frac{dh(x, y_0)}{dx}\Big|_{x=x_0} = \lim_{\varepsilon \to 0} \frac{h(x_0 + \varepsilon, y_0) - h(x_0, y_0)}{\varepsilon}.$$

Problem 61. *Compute the partial derivative over x of the function $z = \sqrt{9 - x^2 - y^2}$ at the point* $(2, 1)$.

> **Solution.** Assigning $y = 1$, we get a function of one variable $z = \sqrt{8 - x^2}$. Its derivative is $-x/\sqrt{8 - x^2}$. For $x = 2$ we obtain
>
> $$\frac{\partial z}{\partial x}(2, 1) = -1.$$

When the point (x_0, y_0) varies, the value $\frac{\partial z}{\partial x}(x_0, y_0)$ becomes a function of the variables x_0 and y_0. Using the normal notations (x, y) instead of (x_0, y_0), one gets a function of the variables x and y denoted by $\partial z/\partial x$ or simply z_x. Thus, for the function $z = \sqrt{9 - x^2 - y^2}$ we obtain

$$\frac{\partial z}{\partial x} = -\frac{x}{\sqrt{9 - x^2 - y^2}}.$$

Once again, to compute the partial derivative z_x, one has to differentiate $z(x, y)$ with respect to x, treating the variable y as an arbitrary constant. The partial derivative with respect to y is defined in the similar way, treating x as a parameter. For the function $z = \sqrt{9 - x^2 - y^2}$ we have

$$\frac{\partial z}{\partial y} = -\frac{y}{\sqrt{9 - x^2 - y^2}}.$$

We also explain the geometric meaning of partial derivatives. Consider the surface in 3-space consisting of all points $(x, y, z(x, y))$ — the *graph* of the given function of two variables. Figure 4 depicts the graph of our favourite function $\sqrt{9 - x^2 - y^2}$.

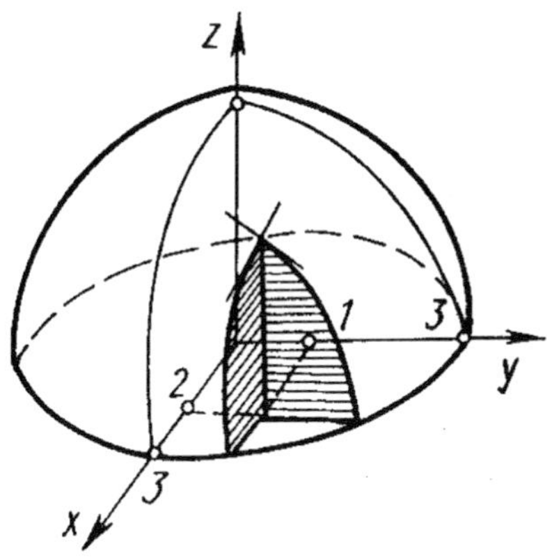

Figure 4. Partial derivatives

Given a point (x_0, y_0), at which the function $z(x, y)$ is defined, draw the plane $y = y_0$. It cuts the surface along a certain plane curve. The slope of the tangent line to this curve at the point (x_0, y_0) gives the value of the partial derivative $z_x(x_0, y_0)$.

Another partial derivative $z_y(x_0, y_0)$ is the slope of the tangent line to the section of the surface by the plane $x = x_0$. Figure 4 shows both sections and their tangent lines for the function $\sqrt{9 - x^2 - y^2}$ at the point $(2, 1)$.

The plane passing through the two tangent lines is the tangent plane to the graph of the function at the given point. Its equation is

$$z - z_0 = p(x - x_0) + q(y - y_0), \tag{67}$$

where $z_0 = h(x_0, y_0)$, $p = \frac{\partial h}{\partial x}(x_0, y_0)$ and $q = \frac{\partial h}{\partial y}(x_0, y_0)$. Indeed, substituting $x = x_0$ into (67), we get the equation of the tangent line to the graph of $h(x_0, y)$ viewed as a function of y, and substituting $y = y_0$, the similar equation for $h(x, y_0)$.

The point of the surface and the point of the tangent plane, both corresponding to one and the same point (x, y) of the horizontal plane, are very close to each other, if the point (x, y) is close enough to (x_0, y_0). Therefore, the difference $z - z_0$, computed by formula (67), can be viewed as an increment of the function $h(x, y)$, when its argument moves from (x_0, y_0) to (x, y), provided that this shift is "infinitely small". Denoting the infinitesimal increments of the three variables by dx, dy and dz, we can write

$$dz = z_x dx + z_y dy \tag{68}$$

(the formula of the *differential of a function of two variables*).

With the help of this formula, we will now obtain the transformation rule for the derivative dy/dx when the variables x and y undergo an arbitrary point transformation

$$\begin{cases} x = \varphi(u, v), \\ y = \psi(u, v). \end{cases} \tag{69}$$

Using the notation $y' = dy/dx$, $v' = dv/du$ (note that the prime has different meaning in either case), by formula (68) we can write

$$y' = \frac{dy}{dx} = \frac{y_u du + y_v dv}{x_u du + x_v dv} = \frac{y_u + y_v \frac{dv}{du}}{x_u + x_v \frac{dv}{du}} = \frac{y_u + y_v v'}{x_u + x_v v'}. \tag{70}$$

We see that the derivative y' is expressed as a fractional linear function of v' with coefficients depending on u and v, i.e., as a projective transformation of v' (see (2)). This remarkable fact, by the way, leads to a deep connection between projective geometry and ordinary differential equations, but we will not discuss that in this book.

As we noticed before, (69) can be viewed either as the passage from one coordinate system in the plane to another, or as a mapping of the plane into itself according to the rule $(u, v) \mapsto (x, y)$. In the latter case (70) describes how the slope of the plane curves changes under this mapping (see Figure 5): v' is the tangent of the angle to the horizontal line for a given curve, while y' is the tangent of the similar angle for the image of this curve.

Let us call a point of the plane together with a direction (a straight line) attached to this point a *contact element*. A contact element is described by three numbers (x, y, p), where (x, y) are the

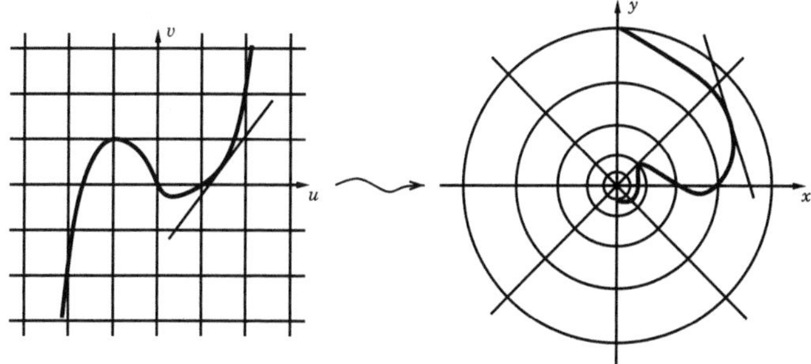

Figure 5. Point transformation

coordinates of the given point and p is the slope of the line (the tangent of the angle it makes with the horizontal axis). The set of all contact elements thus forms a three-dimensional space — the *space of contact elements*. Some of its elements are shown in Figure 6.

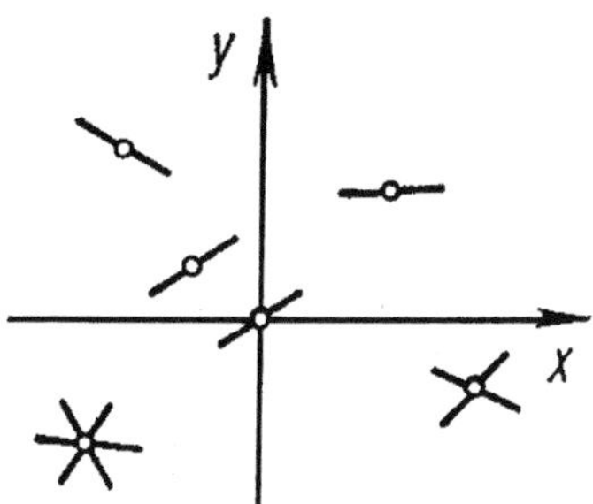

Figure 6. Contact elements

Formula (70), taken together with (69), defines a certain transformation of the space of contact elements, which corresponds to the plane transformation given by formula (69) alone.

Problem 62. *Find the transformation of the space of contact elements that corresponds to the inversion with respect to the circle* $x^2 + y^2 = 1$.

Solution. Under the inversion, the coordinates of a point transform as follows:

$$u = \frac{x}{x^2+y^2},$$
$$v = \frac{y}{x^2+y^2}.$$

The partial derivatives of these functions are

$$u_x = \frac{y^2-x^2}{(x^2+y^2)^2}, \quad u_y = \frac{-2xy}{(x^2+y^2)^2},$$
$$v_x = \frac{-2xy}{(x^2+y^2)^2}, \quad v_y = \frac{x^2-y^2}{(x^2+y^2)^2}.$$

By (70) we obtain

(71) $$v' = \frac{v_x + v_y y'}{u_x + u_y y'} = \frac{(x^2-y^2)y' - 2xy}{-2xyy' + y^2 - x^2}.$$

As a corollary of this result, we can prove the following theorem, which generalizes the assertion of Exercise 153 (see page 187).

Theorem 18. *A plane inversion preserves the angles between curves.*

Proof. The angle between two curves is, by definition, the angle between their tangent lines. If we have two curves in the (x, y)-plane such that the tangents of the slope angles are p_1 and p_2, then the angle α between the curves satisfies

$$\tan\alpha = \frac{p_1 - p_2}{1 + p_1 p_2}.$$

Let q_1 and q_2 be the corresponding tangents for the images of the two curves after inversion, and let β be the angle between them. Then

(72) $$\tan\beta = \frac{q_1 - q_2}{1 + q_1 q_2}.$$

According to (71), we have

$$q_1 = \frac{ap_1 + b}{bp_1 - a},$$
$$q_2 = \frac{ap_2 + b}{bp_2 - a},$$

where a and b are certain constants depending on the point (x, y). Feeding this into (72), we get

$$\tan\beta = \frac{\dfrac{ap_1+b}{bp_1-a} - \dfrac{ap_2+b}{bp_2-a}}{1+\dfrac{ap_1+b}{bp_1-a}\dfrac{ap_2+b}{bp_2-a}} = -\frac{p_1-p_2}{1+p_1p_2} = -\tan\alpha,$$

which means that the inversion preserves the angles, but changes the orientation of the plane. □

We now give an example where a generic point transformation is used to solve a differential equation.

Problem 63. *Solve the differential equation*

$$\frac{(y^2-x^2)\left(x^2+(x^2+y^2)^2\right)+2\,xy\left(y^2+(x^2+y^2)^2\right)}{2\,xy\left(x^2+(x^2+y^2)^2\right)+(x^2-y^2)\left(y^2+(x^2+y^2)^2\right)}$$

Solution. Let us make the change of variables

$$\begin{aligned} x &= \frac{u}{u^2+v^2}, \\ y &= \frac{v}{u^2+v^2}, \\ y' &= \frac{(u^2-v^2)v'-2uv}{-2uvv'+v^2-u^2}. \end{aligned}$$

Substituting these expressions in the given equation, we obtain, after simplifications,

$$v' = \frac{u^2+1}{v^2-1}.$$

Here the variables separate: $(v^2-1)dv = (u^2+1)du$, and the general solution is given by the following implicit function:

$$\frac{v^3}{3} = \frac{u^3}{3} + C.$$

Going back to the variables (x, y), we obtain the answer:

$$y^3 - x^3 = 3(x+y)(x^2+y^2)^2 + C(x^2+y^2)^3,$$

C being an arbitrary constant.

Exercise 172. (a) Find the expression of $dr/d\varphi$ in terms of dy/dx, if (x, y) are Cartesian coordinates and (r, φ) are polar coordinates in the plane. (b) Using the formulas obtained in (a), solve the differential equation $yy' + x = (x^2 + y^2)(xy' - y)$ by passing to polar coordinates.

5. One-parameter groups

Definition 44. A *one-parameter group of plane transformations* is an action of the additive group $\mathbb{R}$ on the plane.

This means that for every real number t a transformation g_t is defined in such a way that the equality $g_t \circ g_s = g_{s+t}$ holds for every pair $s, t \in \mathbb{R}$. In other words, we deal with a homomorphism from the group $\mathbb{R}$ into the group of plane transformations. In this case we say that $\{g_t\}$ is a *one-parameter group of transformations* of the plane. Let us stress that a one-parameter group is not simply the set $\{g_t\}$, but this set together with the parametrization $t \mapsto g_t$.

The simplest example of a non-trivial one-parameter group is the group of parallel translations, say, in the direction of the axis Ox: g_t is the translation by $t\mathbf{e}_1$, where $\mathbf{e}_1$ is the horizontal unit vector.

Exercise 173. Prove that any two transformations belonging to a one-parameter group commute.

A one-parameter group can be written in coordinates as a pair of functions of three variables:

$$\left\{ \begin{array}{lcl} x_t & = & \varphi(x, y, t), \\ y_t & = & \psi(x, y, t). \end{array} \right. \tag{73}$$

Here (x, y) are the coordinates of an arbitrary point in the plane, and (x_t, y_t) are the coordinates of its image under the transformation g_t. For any fixed value of t we obtain a pair of functions of two variables that define a concrete transformation.

The group law, i.e., the relation $g_t \circ g_s = g_{s+t}$, can be written in terms of functions φ, ψ as follows:

$$\left\{ \begin{array}{l} \varphi(\varphi(x, y, s), \psi(x, y, s), t) = \varphi(x, y, s + t), \\ \psi(\varphi(x, y, s), \psi(x, y, s), t) = \psi(x, y, s + t). \end{array} \right. \tag{74}$$

This is the definition of a one-parameter group written in coordinates.

For example, the group of horizontal translations is represented by the functions

$$\begin{cases} x_t = x + t, \\ y_t = y, \end{cases}$$

for which the relations (74) are obviously fulfilled.

Exercise 174. Consider the set of all homotheties with a common centre and positive coefficients. Is this a one-parameter group?

The observant reader will notice that the question of this exercise is not correctly posed, because a one-parameter group presupposes a fixed parametrization of the given set of transformations. If we assign, to every number $t \in \mathbb{R}$, the homothety with stretching coefficient t, we won't obtain a one-parameter group, because the composition of homotheties with coefficients s and t is the homothety with coefficient st, not $s + t$. Fortunately, we know the trick that turns addition into multiplication: this is the exponential function. Assigning the homothety with coefficient e^t to the number t, we get a genuine one-parameter group. Placing the centre of homotheties at the origin, we can describe the group by the functions

$$\begin{cases} x_t = e^t x, \\ y_t = e^t y. \end{cases}$$

Relations (74) obviously hold.

A one-parameter group can be visualized through the set of its orbits. Figure 7 shows the orbits of the two groups that we have mentioned: translations and homotheties.

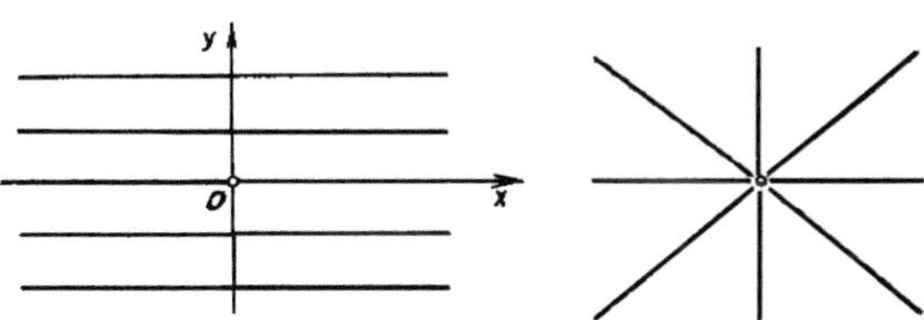

Figure 7. Orbits of one-parameter groups of translations and homotheties

We must note that the set of orbits does not uniquely define the one-parameter group. A simple example of this is provided by the group of translations with a double speed

$$\begin{cases} x_t &= x+2t, \\ y_t &= y, \end{cases}$$

which has the same orbits as the group of translations previously discussed (shown in Figure 7a).

Exercise 175. Define a one-parameter group of rotations with a common centre, write its coordinate representation, and draw its orbits.

Exercise 176. Check that the relations

$$\begin{cases} x_t &= e^{at}(x\cos bt - y\sin bt), \\ y_t &= e^{at}(x\sin bt + y\cos bt) \end{cases}$$

define a one-parameter group. Explain its geometric meaning and draw its orbits.

Exercise 177. Let x_t and y_t be the roots of the quadratic equation for the unknown w

$$(w-x)(w-y)+t=0,$$

chosen in such a way that x_t, y_t continuously depend on t and $x_0 = x$, $y_0 = y$.

The numbers x_t and y_t are functions of three variables x, y and t. Prove that these functions define a one-parameter group of transformations, and draw its orbits.

6. Symmetries of differential equations

A differential equation, viewed as a field of directions in the plane, may possess some symmetry. One glance at Figure 1b is sufficient to understand that this field of directions is preserved by any translation along Oy as well as translations along Ox by whole multiples of 2π. Transformations of the first kind form a one-parameter group $x_t = x$, $y_t = y+t$. Transformations of the second kind form an infinite cyclic group (see p. 87).

It turns out that *the knowledge of a one-parameter group of symmetries enables one to find the general solution of the equation under study in closed form.*

Passing to exact definitions, suppose that we are given a differential equation

$$y' = f(x, y) \tag{75}$$

and a transformation of the plane

$$\begin{cases} x &= \varphi(u, v), \\ y &= \psi(u, v). \end{cases} \tag{76}$$

By (70), we can find the corresponding expression of $y' = dy/dx$ through u, v and $v' = dv/du$.

Definition 45. Transformation (76) is called a *symmetry of the differential equation* (75), if the equation for $v(u)$ obtained after the expressions for u, v and v' are substituted into (75) has the same function f in its right-hand side:

$$v' = f(u, v).$$

In geometric language this means that the transformation of the space of contact elements given by (69) and (70) preserves the surface in this space that consists of all contact elements belonging to the given field of directions.

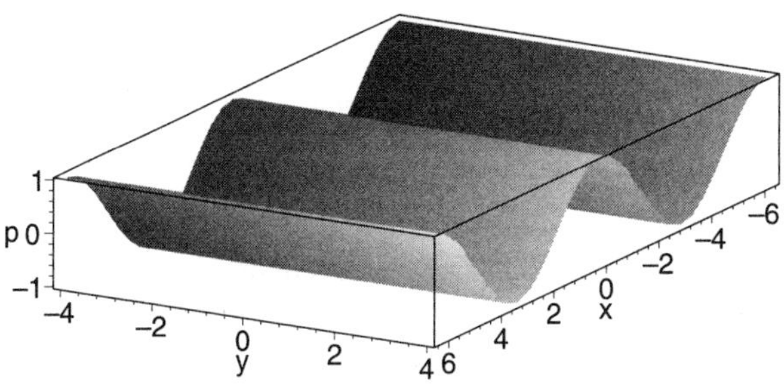

Figure 8. Differential equation as a surface in 3-space

Figure 8 shows such a surface for the differential equation $y' = \cos x$, and you can see that the transformations

$$\begin{aligned} x_t &= x, \\ y_t &= y + t, \\ p_t &= p \end{aligned}$$

and

$$\begin{aligned} x_t &= x + 2\pi k, \\ y_t &= y, \\ p_t &= p \end{aligned}$$

map this surface into itself.

There are two main problems about the interrelation of differential equations and one-parameter groups:

(1) Given a differential equation, find all (or some) groups of its symmetries.

(2) Given a one-parameter group of plane transformations, find all (or some) differential equations preserved by this group, i.e., such that the group consists of their symmetries.

For practical needs (solving differential equations) the first question is more important. But it is also more difficult. Therefore, let us first discuss the second question.

We start with two simple examples, where the answer is obvious:

- The general equation preserved by the group of x-translations is

(77) $$y' = f(y).$$

  We have earlier considered a particular case of this, equation (52).

- The general equation preserved by the group of y-translations is

$$y' = f(x).$$

  We have encountered equation (51), belonging to this class.

Let us now consider more interesting groups.

Problem 64. *Find the general form of a differential equation preserved by the group of rotations of the* (x, y) *plane centred at* $(0, 0)$.

Solution. Under the rotation through an angle α every contact element moves together with the point of attachment and turns by the same angle α (Figure 9).

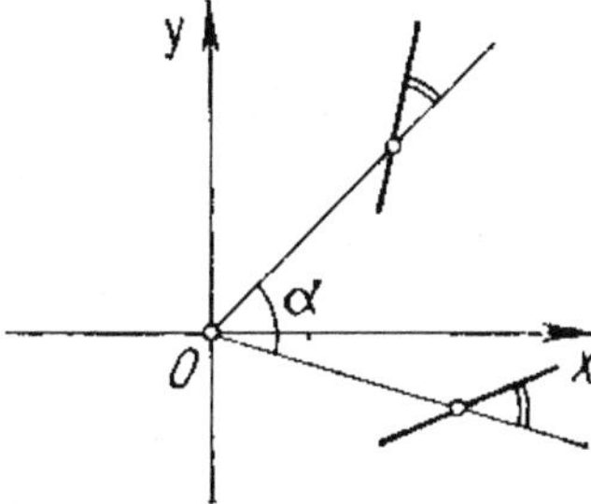

Figure 9. Action of rotations on contact elements

Hence, the angle it makes with the radius-vector of the point (x, y) does not change. Therefore, a field of directions (= a differential equation) is invariant under the group of rotations, if and only if the angle between the direction of the field and the radius-vector depends only on the distance from the origin. We can take the tangent of the angle instead of the angle itself, and the square of the distance instead of the distance. Using the formula for the tangent of the difference of two angles, we obtain the general form of the equation admitting the group of rotations around the origin:

$$\frac{xy' - y}{yy' + x} = f(x^2 + y^2),$$

f being an arbitrary function of its argument. Resolving this with respect to y', we can write the answer as follows:

(78)
$$y' = \frac{xf(x^2 + y^2) + y}{x - yf(x^2 + y^2)}.$$

Exercise 178. Find all differential equations that admit the one-parameter group of homotheties.

Exercise 179. Find all differential equations that admit the one-parameter group of spiral homotheties described in Exercise 176.

7. Solving equations by symmetries

In this section, we will prove the following fact: *if a one-parameter group of symmetries of a differential equation is known, then it can be reduced, by a change of variables, to an equation with separating variables — and hence its general solution can be found in closed form.*

To find the new coordinate system in which the variables separate, we will use *invariants* of one-parameter groups, so let us first mention some of their properties and consider some examples.

We recall (see Section 6) that an invariant of a group action is a function which is constant on the orbits. In other words, a function is an invariant, if it has equal values at any two points that map into each other by a transformation of the group. To give a simple example, any function of y is an invariant of the one-parameter group of translations along Ox. The function y itself is the universal (complete) invariant of this group action, because its values on all orbits are different.

In the same way, the function x is the universal invariant of the group of translations in the direction of the axis Oy.

What is the universal invariant of the group of homotheties with centre 0 acting on the plane without the origin? One is tempted to think that it is the polar angle φ. Indeed, the polar angle φ takes equal values at all points of every ray (half-line) that issues from the origin, and different rays correspond to different values of the function. However, the polar angle φ is not a normal single-valued function on the plane, for example, to the point $(-1, 0)$ one can, with equal truth, assign the values $180°$ and $-180°$ (and an infinite number of others). Of course, one can make φ a single-valued function using, for example, the convention that φ must always take values between $0°$ (including) and $360°$ (excluding) — but then it will become a discontinuous function. In fact, the group of homotheties does not have any continuous universal invariant with values in $\mathbb{R}$. Of course, it does have continuous invariants which are not universal, for example, the function $\sin\varphi$.

Exercise 180. Find an invariant of the group of rotations around the origin. Does this group have a universal continuous real-valued invariant?

There are two approaches to the problem of finding the invariants of a given one-parameter group:

(1) *The formal approach.* Assuming that the group is given in coordinates by a pair of functions (73), the problem is to make up a combination of the expressions $\varphi(x, y, t)$ and $\psi(x, y, t)$ that does not contain the variable t, i.e., to find a function $h(\varphi, \psi)$ that does not depend on t.

(2) *The geometric approach.* In the plane, we draw a curve K that meets every orbit of the group in exactly one point (see Figure 10). We choose an arbitrary function on this curve that takes different values at different points, and then prolong it to the whole plane, following the rule that the value of the function at any point A is set equal to its value at the point B where the orbit meets the chosen curve K.

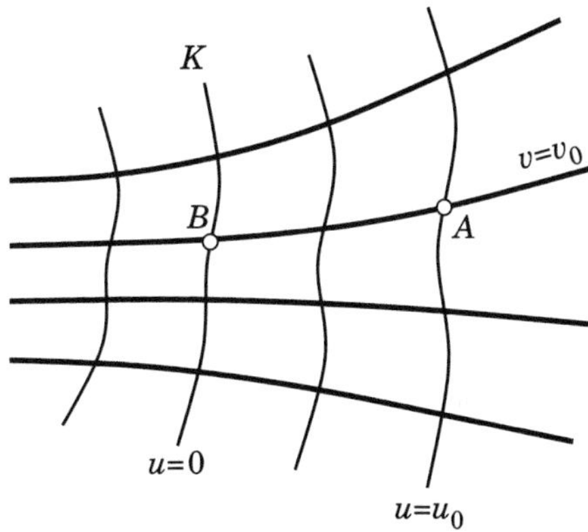

Figure 10. Invariant of a one-parameter group

Problem 65. *Find an invariant of the one-parameter group*

$$\begin{cases} x_t = x + t, \\ y_t = e^t y. \end{cases}$$

Solution. *Formal method.* Looking at the above formulas for a while, one can guess that the combination $e^t y \cdot e^{x+t} = ye^{-x}$ does not depend on t and thus provides an invariant of the group.

Geometric method. The orbits of the given group are shown in Figure 11.

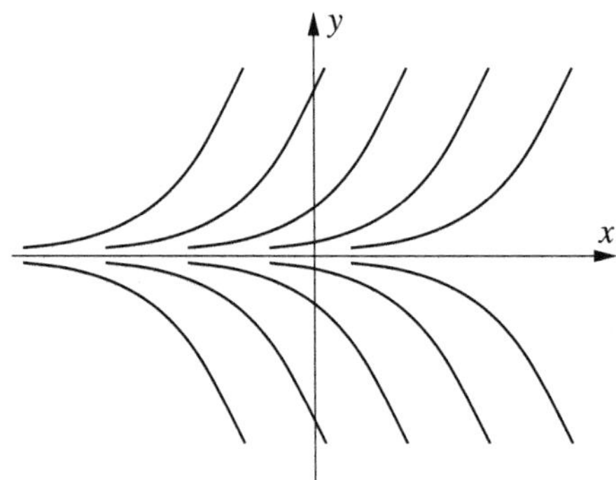

Figure 11. Orbits of the group studied in Problem 65

The coordinate axis Oy meets every orbit in one point, so we can choose it as the curve K. The function $v(0, y) = y$ takes different values at different points of this line, so let us find its prolongation to all the plane along the orbits of the group. Let $A(x, y)$ be an arbitrary point of the plane and let $B(0, v)$ be the intersection point of the corresponding orbit with the y-axis. From the equations of the group we see that there is a number t such that $x = t$, $y = e^t v$. From these equations we find that $v = ye^{-x}$. Thus, the invariant is ye^{-x}.

Exercise 181. Find a non-trivial invariant of the group of spiral homotheties (Exercise 176) by the above two methods.

We finally get to the question of integrating differential equations with a known symmetry group. The first principle is that it is useful to pass to new coordinates such that one of the new coordinate functions is an invariant of the group. After such transformation it often (but not always) happens that the variables separate, and the equation can be solved in closed form. Let us consider an example that comes up quite often.

Problem 66. *Find the general solution of the differential equation*

$$y' = f(y/x). \tag{79}$$

Solution. Such equations are called *homogeneous*. As we know from Exercise 178, homogeneous equations are invariant with respect to the one-parameter group of homotheties with the centre $(0, 0)$. Since the function y/x is an invariant of this group, let us take it for the new dependent variable, leaving the independent variable x unchanged. We thus set $v = y/x$, or $y = xv$; hence $y' = v + xv'$. Substituting this into (79), we get $v + xv' = f(v)$, or

$$v' = \frac{f(v) - v}{x}.$$

This is indeed an equation with separating variables.

Exercise 182. Continuing the previous argument, find an explicit answer in the particular case of the equation $y' = 1 + 2y/x$.

Exercise 183. Do the variables in (79) separate in polar coordinates?

Exercise 184. Adapt the argument of Problem 66 to the equations of the form

$$y' = f\left(\frac{ax + by + c}{a_1x + b_1y + c_1}\right).$$

We must stress that taking the new dependent variables to be an invariant of the group does not guarantee the separation of variables. The choice of the *independent* variable is also very important. For example, if, in the equation (78) that admits the group of rotations, one goes over to polar coordinates, then the new equation has the form $u\,dv/du = f(v)$, where the variables separate. However, the transformation $u = x$, $v = x^2 + y^2$ (the second variable is a group invariant) does not lead to an equation with separating variables (please check).

To get a universal rule for the integration of equations with a known one-parameter group of symmetries, we will try to find a coordinate system (u, v) such that in these coordinates the group looks as simple as possible, for example, consists of parallel translations along the axis u. The general form of an equation that admit this group is, as we know, $dv/du = \varphi(v)$. Here the variables separate, and the equation is solvable in closed form. It remains to understand how we can reduce the initial one-parameter group to this simple form by a change of coordinates.

In the coordinates (u, v) we want the transformations of the group to have the form $u_t = u + t$, $v_t = v$. These transformations map the coordinate line $u = 0$ into the parallel lines $u = t$. Choose a real-valued function $v(x, y)$ which is an invariant of the group. The orbits are given by the equation $v = \text{const}$. Choose a curve K which meets every orbit in one point, and assume that K is the v-axis of the new coordinate system, i.e., is described by the equation $u = 0$. Then, to fulfill our plane, we must assume that the image of K under the group transformation g_t should be described by the equation $u = t$.

If, instead of this function u, we take for the independent variable another function w that has the same level lines $w = \text{const}$ (the line $w = t$ should be the image of $w = 0$ under some transformation of the group, but not necessarily g_t), then we will also have an equation with separating variables in the new coordinates. Indeed, in this case w is a function of u: $w = h(u)$, which, upon substitution into the equation $dv/du = \varphi(v)$, gives an equation $dv/dw = \varphi(v)\psi(w)$.

We can state this result as the following *variable separation theorem.*

Theorem 19. *Suppose that we know a one-parameter group $G = \{g_t\}$ of symmetries of a differential equation E. Then the equation E becomes an equation with separating variables in any coordinate system (u, v) such that the coordinate lines $v = \text{const}$ are the orbits of the group G, and the lines $u = \text{const}$ go into each other under the transformations g_t.*

Problem 67. *Solve the differential equation*

$$y' = \frac{2}{5}(y^2 + x^{-2}).$$

Solution. If x is multiplied by a constant k and y by its inverse k^{-1}, then y' gets multiplied by k^{-2} (you can also check that using formula (70)). All the terms of the given equation increase by the same factor, so that the equation actually does not change. This means that the group

$$\begin{cases} x_t &= e^{-t}x, \\ y_t &= e^t y \end{cases}$$

is a group of symmetries of the given equation.

This group is called the group of *hyperbolic rotations.* Its orbits are branches (connected components) of the hyperbolas $xy = \text{const}$ (see Figure 12). The product xy is an invariant of this group. We take it as the new dependent variable: $v = xy$.

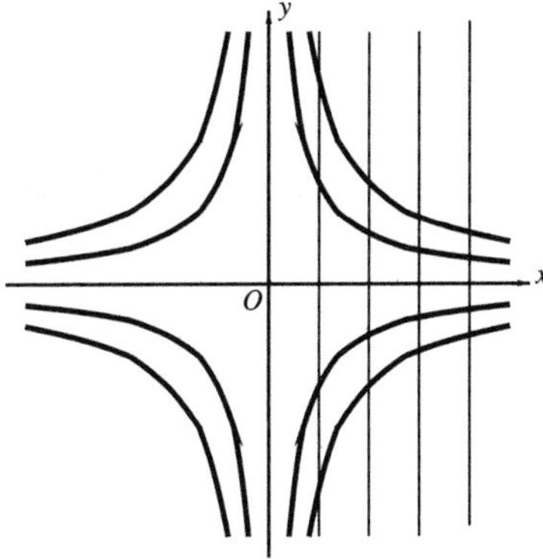

Figure 12. Orbits of hyperbolic rotations

Note that the vertical lines $x = \text{const}$ are mapped one into another by hyperbolic rotations: the image of the line $x = a$ is the line $x = e^{-t}a$. Therefore, we can set x to be the new coordinate u. The required change of variables is thus

$$\begin{aligned} x &= u, \\ y &= \frac{v}{u}. \end{aligned}$$

After this change of variables the given equation becomes

$$v' = \frac{2v^2 + 5v + 2}{5u}.$$

The variables separate, and the general integral is

$$\frac{v + \frac{1}{2}}{v + 2} = Cu^{3/5}.$$

Expressing u and v through x and y, we obtain the answer:

$$y = \frac{3}{2(x - Cx^{8/5})} - \frac{2}{x}.$$

Remark. The same procedure solves the equation $y' = a(y^2 + x^{-2})$ for an arbitrary value of a. The reason why we only considered a special case is that the general answer is rather cumbersome.

Exercise 185. Find the general solution of $y' = (x + y^2)/(xy)$.

As the last example, we consider an equation where it is impossible to separate the variables by a change of the form $x = \varphi(u)$, $y = \psi(x, y)$.

Problem 68. *Solve the differential equation*

$$y' = \frac{(y^2 - x^2)^2 - 5(y - x) + 4}{(y^2 - x^2)^2 + 5(y - x) + 4}$$

using the one-parameter group

$$\begin{cases} x_t &= \frac{e^t + e^{-t}}{2} x + \frac{e^t - e^{-t}}{2} y, \\ y_t &= \frac{e^t - e^{-t}}{2} x + \frac{e^t + e^{-t}}{2} y. \end{cases}$$

Solution. These transformations are hyperbolic rotations (see 7), but considered in a coordinate system rotated through 45° with respect to the initial one. The function $v = y^2 - x^2$ is a group invariant. Note that the families of vertical or horizontal lines are not preserved by the transformations of the group; hence it is no good to take $u = x$ or $u = y$. However, the lines $y - x = \text{const}$

do have this property. Set $u = y - x$. Then after the transformation the equation becomes

$$\frac{dv}{du} = \frac{5v - v^2 - 4}{u}.$$

The variables separate!

Solving this equation and making the inverse change of variables, we obtain the implicit solution to the initial equation:

$$\frac{y^2 - x^2 - 1}{y^2 - x^2 - 4} = C(y - x)^3.$$

Exercise 186. Solve the differential equation $y' = e^{-x}y^2 - y + e^x$ with the help of the symmetry group $x_t = x + t$, $y_t = e^t y$.

We have thus learned how to solve a differential equation, if a one-parameter group of its symmetries is known. "This, however, by no means implies that any differential equation $X dy - Y dx = 0$ can be solved in closed form. The difficulty consists in *finding* the one-parameter group that leaves it invariant". These words were written by Sophus Lie, the Norwegian mathematician who created the theory of continuous groups and found its applications to differential equations, the simplest case of which was described in the previous pages. As a last remark, let us mention that during the last 30 years or so some algorithmic methods for finding the symmetries of differential equations have been elaborated and implemented in numerous software systems of computer algebra.

Answers, Hints and Solutions to Exercises

1. Example: triangle with vertices $(0,0)$, $(12,9)$, $(24,-7)$.
2. The problem can be solved either by a direct construction or using the result of exercise 6. Answer: the polygon Φ may have 3, 4, 5 or 6 vertices.
3. A necessary and sufficient condition is $k - l = 1$. To prove this, rewrite the given expression as $A_1 + A_2 - B_1 + A_3 - B_2 + \cdots$.
4. Problem 2 shows that the given assertion is equivalent to *Euler's theorem*, which says that $\overrightarrow{MH} = 2\overrightarrow{OM}$, where M is the median intersection point, O is the outcentre (the centre of the circumscribed circle) and H is the orthocentre (the intersection point of the altitudes) of the triangle. To prove this theorem, construct the triangle $A_1B_1C_1$ which is twice as large as ABC and with sides parallel to those of ABC (see Figure 1) and observe that the perpendicular bisectors of the sides of triangle ABC coincide with the altitudes of triangle $A_1B_1C_1$.
5. The necessary and sufficient condition is $\alpha+\beta+\cdots+\omega = 1$. The proof follows from the vector equality

$$\alpha\overrightarrow{PA} + \beta\overrightarrow{PB} + \cdots + \omega\overrightarrow{PZ} = \alpha\overrightarrow{QA} + \beta\overrightarrow{QB} + \cdots + \omega\overrightarrow{QZ} + (\alpha+\beta+\cdots+\omega)\overrightarrow{PQ}.$$

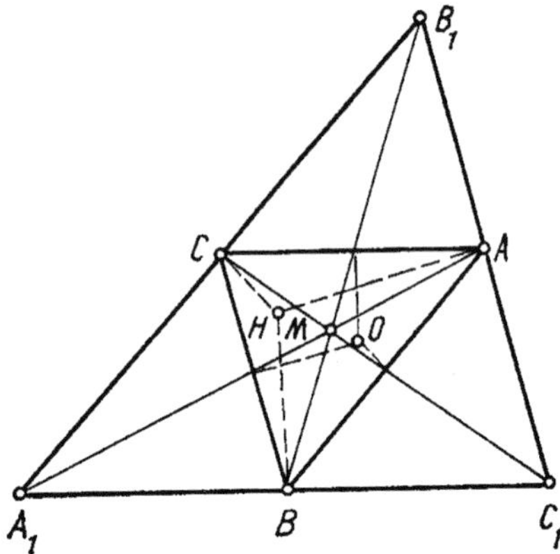

Figure 1. Euler's theorem

6. The set of all points $a_1A_1 + a_2A_2 + \cdots + a_nA_n$ where all coefficients a_i are nonnegative and $a_1 + a_2 + \cdots + a_n = 1$.
7. Express all the points under consideration in terms of the four vertices of the quadrangle.
8. Express these points in terms of the vertices of the hexagon.
9. Express all the points under consideration in terms of the four vertices of the quadrangle.
10. Take the intersection point of the two medians for the pole and use the result of Problem 2.
11. Answer: the diagonal is divided in the ratio 1:6. Hint: take one vertex of the parallelogram for the pole, two others for the basic points, and express the intersection point of the given line and the diagonal in two different ways.
12. (a) $y = b$, (b) $x = a$, (c) $ay = bx$.
13. The union of the three medians.
14. Let K be the pole and A, B, the basic points. Express the coordinates of the points D, E, F in terms of the coordinates (a, b) of the point C.
15. See the solution to the previous exercise.

16. To both questions the answer is negative.

17. Expand the point $K(0,1)$ in terms of the basis E, A and prove that $K^2 = -E$. Therefore, this multiplication coincides with the multiplication of complex numbers (see p. 30). Using the trigonometric representation of complex numbers, we obtain the following answer:

	E	A	B	C	D
E	E	A	B	C	D
A	A	B	C	D	E
B	B	C	D	E	A
C	C	D	E	A	B
D	D	E	A	B	C

18. (a) 0, (b) $\pm(2-i)$, (c) 1 (note that the cube of the given number is -1).

19. (a) Circle of radius 5 centred at point -3. (b) Perpendicular bisector of the segment $[-4, 2i]$. (c) Circle (use the identity $|p|^2 + |q|^2 = (|p+q|^2 + |p-q|^2)/2$ for complex numbers p and q).

20. The sum in the left-hand side of the equality is equal to the length of a certain broken line connecting the points 0 and $5+5i$ of the complex plane. To see this, observe that the sum of the numbers $x_1 + (1-x_2)i$, $(1-x_3) + x_2 i$, ..., $x_9 + (1-x_{10})i$, $(1-x_1) + x_{10}i$ is $5+5i$.

21. $0°$, $90°$, $180°$, $270°$, $45°$, $330°$.

22. (a) See Figure 2.
(b) For example, $(r-2)(r-2-|\sin 3\phi|) = 0$ or, in Cartesian coordinates,

$$(\sqrt{x^2+y^2}-2)(\sqrt{x^2+y^2}-2-\frac{|3x^2y-y^3|}{(x^2+y^2)^{3/2}}) = 0.$$

23. Prove that $z = \cos\alpha \pm i\sin\alpha$.

24. Answer: $2n$. Proof: similar to Problem 9.

25. If the point z lies outside of the polygon, then the arguments of all differences $z - a_i$ belong to one segment of width $180°$. Therefore, the arguments of all numbers $1/(z-a_i)$ also lie in

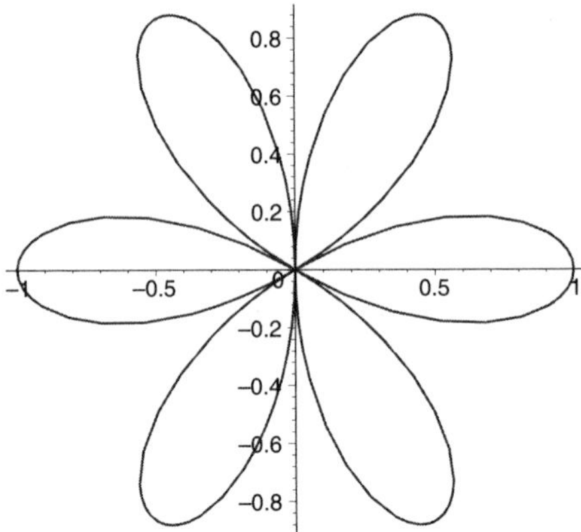

Figure 2. The curve of Exercise 22(a)

one segment of the same width, and these numbers cannot sum up to 0.

26. Use two translations.

27. One possible solution is to draw the lines parallel to the sides of the given triangle, through the endpoints of the given vector, thus *circumscribing* the triangle around the vector, and then make a parallel translation to move the triangle to its initial position.

28. Let the trapezoid be $ABCD$ with parallel sides AB and CD. Translate the point B by vector $\overrightarrow{DC}$ to a point B' and consider the triangle ACB'.

29. Use two reflections in the sides of the given angle.

30. The point symmetric to a vertex of the triangle with respect to a bisector lies on the opposite side of the triangle (or its prolongation).

31. Roll the angle by successive reflections four times and unfold the reflections of the ray. The phenomenon occurs for all angles $90^\circ/n$ with integer n.

32. After the rotation through 90° the vectors $\overrightarrow{MA_i}$ become the consecutive sides of the given polygon.

33. Use rotations through 60°.

34. Let K be the result of rotation of the point M around the centre of the square which takes vertex A into B. Show that $AK \perp BM$, $BK \perp CM$, $CK \perp DM$, $DK \perp AM$.

35. Use the point symmetry with respect to the intersection point of the circles.

36. The first player should put the first coin in the centre of the table, then use point symmetry.

37. Similar to the proof of formula (7) (p. 52). Use a rotation around the intersection point of the given line with the x-axis.

38. Introduce a suitable complex structure in the plane.

39. Answer: the identity. Proof: use formula (12) (p. 56).

40. Given three numbers that correspond to the points A, B, C, by formula (17) find the numbers that correspond to M, N and P.

41. It is easy to check that $R^d_M \circ R^d_N \circ R^d_P \circ R^d_Q = \mathrm{id}$. Rewrite this condition in terms of complex numbers.

42. (a) $R_B \circ R_A = T_{2\overrightarrow{AB}}$, (b) $S_l \circ R_A$ is a glide reflection with axis AK and vector $2\overrightarrow{AK}$, where K is the base of the perpendicular drawn from the point A to the line l.

43. Consider the composition of five point symmetries in the centres of the consecutive sides of the pentagon. Prove that it is a point symmetry with respect to one of the vertices of the pentagon.

44. Find the axis of the glide symmetry which is the composition of the three given reflections. Consider the line that passes through the given point parallel to this axis.

45. Use the result of Exercise 37.

46. No, because each shot changes the orientation of the set of three pucks.

47. Use the definition of the argument.

48. (a) $R_A \circ S_l = S_l \circ R_A$, (b) $R_B \circ R_A = R_A \circ R_C$.

49. (a) The line l is the perpendicular bisector of the segment AB, (b) The lines l, m and n meet in one point.

50. The set of all points $(5-6k+12l, 3+12k-6l)$, $(7-6k+12l, 2+12k-6l)$, $(3-6k+12l, 1+12k-6l)$, $(4-6k+12l, -1+12k-6l)$, $(2-6k+12l, 9+12k-6l)$, $(3-6k+12l, 10+12k-6l)$, where k and l are arbitrary integers.

51. (a) yes, (b) yes, (c) no.

52. Take an arbitrary element $g \in G$. By property (2), $g^{-1} \in G$. By property (1), $g \circ g^{-1} = \mathrm{id} \in G$.

53. Use Figure 3.1b and the fact that a movement is completely determined by the images of three non-collinear points.

54. All these groups are different.

55. Use the result of Problem 21.

56. (a) Yes, for example, a rectangle, (b) no. If R_A and R_B are symmetries of a certain figure, then R_C with $C = R_A(B)$ is also a symmetry.

57. A circle, in particular a point, or the union of a set of concentric circles.

58. All possible conjugations in the group of plane movements are displayed in the following table, which, on the intersection of a row labelled g and a column labelled f, contains the element $f \circ g \circ f^{-1}$.

	$T_{\mathbf{b}}$	R_B^β	S_m	$U_m^{\mathbf{b}}$
$T_{\mathbf{a}}$	$T_{\mathbf{a}}$	$T_{R_B^\beta(\mathbf{a})}$	$T_{S_m(\mathbf{a})}$	$T_{S_m(\mathbf{a})}$
R_A^α	$R^\alpha_{T_{\mathbf{b}}(A)}$	$R^\alpha_{T_{\mathbf{b}}(A)}$	$R^{-\alpha}_{S_M(A)}$	$R^{-\alpha}_{U_m^{\mathbf{b}}(A)}$
S_l	$S_{T_{\mathbf{b}}(l)}$	$S_{R_B^\beta(l)}$	$S_{S_m(l)}$	$S_{U_m^{\mathbf{b}}(l)}$
$U_l^{\mathbf{a}}$	$U^{\mathbf{a}}_{T_{\mathbf{b}}(l)}$	$U^{R^\beta(\mathbf{a})}_{R_B^\beta(l)}$	$U^{S_m(\mathbf{a})}_{S_m(l)}$	$U^{S_m(\mathbf{a})}_{U_m^{\mathbf{b}}(l)}$

59. The multiplication table for the group D_3 is shown on page 92.

60. The identity transformation commutes with everything. All rotations commute between themselves. Apart from these, in the group D_n with odd n there are no more commuting pairs. In the group D_n with even n we also have commuting pairs of reflections in mutually perpendicular lines.

61. Verify the equality $(f^{-1})^k \circ f^k = \text{id}$.

62. The order of f^k is equal to $n/\operatorname{GCD}(n, k)$.

63. (a)

2	3	4	5	6	7	8	9	10	11	12	13	14	15
1	2	2	4	2	6	4	6	4	10	4	12	6	8

(b) $\varphi(m) = m(1 - 1/p_1) \cdots (1 - 1/p_k)$, where p_1, ..., p_k are all prime divisors of m.

64. No. Some straight lines go into themselves under this movement.

65. The composition of n rotations R_A^α yields an identity if and only if the total angle of rotation, $n\alpha$, is equal to a multiple of 360°.

66. The group generated by A and B is C_5, a cyclic group of order 5. A is a rotation by an angle $\alpha = \pm 72°$ or $\pm 144°$, B is the rotation around the same centre by the angle -2α.

67. No. Any word is equivalent to a word of the form $EI^kA^lU^m$, where k, l, m are integers between 0 and 6.

67. One can take either a suitable pair of reflections or a pair consisting of a reflection and a suitable rotation. For example, two reflections S_1 and S_2 whose axes are adjacent satisfy the defining relations $S_1^2 = S_2^2 = (S_1 \circ S_2)^n = \text{id}$.

69. Start by proving that none of the movements F_i can be a rotation.

70. Under the assumptions of the exercise, prove the following two facts: (1) for any given element a there are two integers m and n such that $a^m = a^n$, (2) the law of cancellation holds: if $xy = xz$, then $y = z$.

71. (1) no: the numbers $\sqrt{2}$ and $-\sqrt{2}$ are irrational, but their sum is rational. (2) yes, (3) yes, (4) no: the inverse of 3/4 is

not binary-rational, (5) for example, the set of all numbers $\{\tan n \mid n \in \mathbb{Z}\}$.

72. $x = ba^{-1}$.

73. Check that this operation is associative.

74. $G = \{x, 1/(1-x), (x-1)/x, 1-x, 1/x, x/(x-1)\}$. Compositions of these functions are given in the following table:

	x	$\frac{1}{1-x}$	$\frac{x-1}{x}$	$1-x$	$\frac{1}{x}$	$\frac{x}{x-1}$
x	x	$\frac{1}{1-x}$	$\frac{x-1}{x}$	$1-x$	$\frac{1}{x}$	$\frac{x}{x-1}$
$\frac{1}{1-x}$	$\frac{1}{1-x}$	$\frac{x-1}{x}$	x	$\frac{1}{x}$	$\frac{x}{x-1}$	$1-x$
$\frac{x-1}{x}$	$\frac{x-1}{x}$	x	$\frac{1}{1-x}$	$\frac{x}{x-1}$	$1-x$	$\frac{1}{x}$
$1-x$	$1-x$	$\frac{x}{x-1}$	$\frac{1}{x}$	x	$\frac{x-1}{x}$	$\frac{1}{1-x}$
$\frac{1}{x}$	$\frac{1}{x}$	$1-x$	$\frac{x}{x-1}$	$\frac{1}{1-x}$	x	$\frac{x-1}{x}$
$\frac{x}{x-1}$	$\frac{x}{x-1}$	$\frac{1}{x}$	$1-x$	$\frac{x-1}{x}$	$\frac{1}{1-x}$	x

75. For example, the sum of squares of all expressions found in Exercise 74.

76. Yes.

77. Elements S_a, S_b, S_c can be arbitrarily permuted. The total number of isomorphisms, including the identity, is 6.

78. A direct verification of the definition of the isomorphism.

79. Point E corresponds to the identity transformation, while the points A, B, C, D, K correspond to rotations by 60°, 120°, 180°, 240°, 300°, respectively.

80. The circle corresponds to the identity transformation, while the triangle and the square correspond to the remaining two elements of the group C_3. There are two different isomorphisms.

81. Write out the two multiplication tables. By permuting the rows and columns in one table, make it look like the other one, up to notation of the elements.

82. Yes. The correspondence $k \leftrightarrow 2k$ is an isomorphism.

83. The only pair of isomorphic groups is D_1 and C_2.

84. Considering the expression $\varphi(g \circ g^{-1})$, prove that $\varphi(g^{-1}) = h^{-1}$. It follows that $\varphi(g^{-n}) = \varphi((g^{-1})^n) = (h^{-1})^n = h^{-n}$.

85. (1) Number the vertices of the equilateral triangle. (2) Consider the action of this group on the extended real line and the permutations of the set $\{0, 1, \infty\}$ under this action.

86. Denoting the Napier logarithm by N, we will have $N(x_1x_2) = N(x_1) + N(x_2) - B$.

87. (a) A direct check of group axioms. (b) The inverse image of this group under the mapping $\varphi(x) = \tan x$ would be an additive group of real numbers containing some open interval. Prove that it coincides with the whole of $\mathbb{R}$.

88. (a) $x \star y = \sqrt[3]{x^3 + y^3}$, (b) this operation is the pullback of multiplication along the mapping $x \mapsto x - 1$.

89. The group D_3 has four proper subgroups: one of order 3 and three of order 2.

90. Yes, the number -1.

91. Prove that any subgroup of $\mathbb{Z}$ is generated by its smallest positive element.

92. Check the group axioms. The notion of the quotient group (section 2) provides an easier way to prove this fact.

93. No, because, for example, $\bar{2} \cdot \bar{3} = \bar{6} = \bar{0}$.

94. Yes. To prove this, write out the multiplication table.

95. Every solution of the equation $x^2 = 3y^2 + 8$ is also a solution of the equation $x^2 \equiv 3y^2 + 8 \pmod 3$, which is equivalent to $x^2 \equiv 2 \pmod 3$. However, this last equation has no solutions.

96. Answer: 81.

    The last two digits of a positive integer is the same thing as its residue modulo 100. The number 2003 is mutually prime with 100, because it is not divisible by 2 and 5. Since $\phi(100) = 100 \cdot (1 - 1/2) \cdot (1 - 1/5) = 40$ and $2004 \bmod 40 = 4$, by Fermat's little theorem we have: $2003^{2004} \equiv 2003^4 \equiv 3^4 = 81$.

97. A homomorphism from $\mathbb{Z}_m$ onto $\mathbb{Z}_n$ exists if and only if m is divisible by n. Under this assumption, one of the possible homomorphisms is given by the correspondence $\bar{a} \mapsto \bar{a}$, where the bar on the left means the residue class modulo m, while the bar on the right means the residue class modulo n.

98. One can get one of the following expressions: x, $1-x$, $1/x$, $1/(1-x)$, $1-1/x$, $x/(x-1)$. Compare this with the result of Exercise 74.

99. Use the classification of plane movements (Theorem 4 in section 7).

100. Prove that $\angle OAC = \angle BEF$, denote this angle by α and consider the rotation of BE around E and the rotation of AC around A through α.

101. Both assertions follow from the fact that the determinant of the product of two matrices equals the product of the determinants of the two matrices.

102. (a) Yes. (b) No.

103. The kernel consists of all odd functions, the image of all even functions.

104. Consider the homomorphism $\varphi(z) = z^n$.

105. Consider the homomorphism $\varphi(x) = \cos x + i \sin x$.

106. This is the dihedral group D_n.

107. If $aba = bab$, then the elements $x = ab$, $y = aba$ satisfy $x^2 = y^3$.
If $x^2 = y^3$, then the elements $a = x^{-1}y$, $b = y^{-1}x^2$ satisfy $aba = bab$.

108. The set splits into 5 orbits.

109. $\{-1, 2, 1/2\}$.

110. $\{1/2 + i\sqrt{3}/2, 1/2 - i\sqrt{3}/2\}$.

111. In both cases the action is transitive. Every edge is preserved by 2 movements, and every vertex is preserved by 3 movements.

112. (a) D, S, T, (b) D.

113. (a) 2, (b) 1, (c) 7.

114. $\displaystyle\frac{1}{m}\sum_{k=1}^{m} n^{\mathrm{GCD}(k,m)}$, where GCD stands for the greatest common divisor.

115. 60.

116. 16.

117. 30.

118. (a) 23, (b) 218.

119. $(\binom{15}{6} + 15\binom{7}{3} + 2\binom{5}{2})/30 = 185$.

120. Here is one construction of a complete invariant; it is not tremendously elegant. but we describe it for want of a better one. The black beads split the set of all white beads into 4 parts, some of which may be empty. Let m be the number of white beads in the biggest part, and n, k the numbers of beads in the two adjacent parts such that $n \geq k$. Assign the triple (m, n, k) to the given necklace. For a necklace with several biggest parts, among all the triples $(m.n.k)$ choose the lexicographically biggest one. (Actually, there is only one necklace for which we must worry about this.) Then the triple (m, n, k) is a complete invariant of the necklace.

121. (a) The well-known criteria for the equality of two triangles (angle and two adjacent sides, side and two adjacent angles, three sides) provide examples of complete invariants.
(b) The ordered set consisting of the lengths of all sides AB, BC, CD, DA and one angle ABC is an invariant. This invariant is complete on the set of convex quadrilaterals.

122. See the discussion of finite rotation groups on page 79.

123. Note that an infinite set of points on a circle cannot be discrete.

124. Angle of $360°/n$ for C_n or $180°/n$ for D_n with vertex at the common centre of rotations.

125. The condition $|kn - lm| = 1$ means that the area of the parallelogram is 1. Therefore, by Pick's well-known formula, if P is such a parallelogram with one vertex at point A, then the points of the orbit of A that belong to P are only the vertices of P.

126. Use the table of conjugations (page 234).

127. Let O be an arbitrary point of the plane, and S the orbit of A under the action of the given group. Let A be an arbitrary point of S such that the segment OA does not contain other points of S. Let B be a point of S at the minimal distance from the line OA. Prove that the pair $\overrightarrow{OA}, \overrightarrow{OB}$ is a system of generators.

128. Group of order 18 with generators a, b, c and relations $a^2 = b^2 = c^2 = (ab)^3 = (bc)^3 = (ca)^3 = (abc)^2 = e$.

129. $p4m$, $p4g$, $p2$.

130.

C_1	$p1$
D_1	pm, pg, cm
C_2	$p2$
D_2	pmm, pmg, pgg, cmm
C_3	$p3$
D_3	$p31m$, $p3m1$
C_4	$p4$
D_4	$p4m$, $p4g$
C_6	$p6$
D_6	$p6m$

131. It is easy to see that a number, if present, stands for the maximal order of a rotation, the symbol m ('*mirror*') is for a reflection, g for a glide reflection. The distinction between p and c is more subtle: the crystallographers' implication is *primitive* or *centred* cell, but that does not seem to have any mathematical meaning other than the groups with a c may have a cell in the form of an arbitrary rhombus (see page 160), which leaves a layman wondering why no similar notation is used for rectangles or hexagons. Finally, the order of m and 1 in the notations $p3m1$ and $p31m$ looks completely enigmatic.

132. Define a homomorphism of $\mathrm{Aff}(2, \mathbb{R})$ onto $\mathrm{GL}(2, \mathbb{R})$ and use the first homomorphism theorem (page 134).

133. There is an affine transformation that takes the given trapezoid to a trapezoid with equal sides.

134. Both groups have order 6 and are isomorphic to the dihedral group D_3.

135. Point m/p, if $p \neq 0$.

136. Write out explicit formulas for the composition and the inverse transformation.

137. By a direct computation, check that

$$\frac{x_3' - x_1'}{x_3' - x_2'} : \frac{x_4' - x_1'}{x_4' - x_2'} = \frac{x_3 - x_1}{x_3 - x_2} : \frac{x_4 - x_1}{x_4 - x_2},$$

if $x_i' = (mx_i + n)/(px_i + q)$.

138. As in Exercise 74, the complete list of elements of this group can be obtained directly, taking the compositions of the two given functions until the elements begin to repeat.

To prove the isomorphism, find two generating elements of this group that satisfy the defining relations of the group D_n (see formula (25) on page 92).

139. The transformation $x \mapsto (mx + n)/(px + q)$ has finite order if and only if

$$\frac{m^2 + q^2 + 2np}{2(mq - np)} = \cos\alpha,$$

where α is an angle measured by a rational number of degrees.

140. No. Using formula (43), find an example of a translation t and a projective transformation p such that ptp^{-1} is not affine.

141. By a projective transformation, the quadrangle $ACFD$ may be made into a square. Then we can introduce a Cartesian coordinate system adapted to this square and solve the problem by a direct computation. *Remark.* Of course, this is not a very beautiful solution. It only shows how to reduce the amount of computations needed to solve the problem by brute force. Indeed, in the original setting we had 6 basic points described in coordinates by a set of 12 real numbers related by two equations. After the transformation, the configuration is described by only two independent parameters.

142. Define a homomorphism of $\mathrm{GL}(2, \mathbb{R})$ onto $\mathrm{PGL}(1, \mathbb{R})$ and use the first homomorphism theorem (page 134).

143. It is easier to construct a triangle, circumscribed around a given triangle, with sides parallel to the three given lines.

144. Use a homothety with centre on the outer circle and coefficient $3/5$.

145. Use a homothety with centre at the median intersection point and coefficient -2.

146. Consider the homothety with coefficient $1/2$ centred at the intersection point of the three altitudes. Prove that the outcircle of the given triangle goes into the desired circle under this transformation.

147. We know that any similitude with coefficient different from 1 has a unique fixed point. The problem is to prove that this point lies inside the smaller map. Suppose that it lies outside, draw a line through this point that intersects the smaller map, and consider its intersection points with the boundaries of the maps.

148. E is the centre of homothety that takes A into B and C into D.

149. Find the images of segments MC and PN under the action of the spiral homotheties $F_A(\sqrt{2}, 45°)$ and $F_C(\sqrt{2}, 45°)$, respectively.

150. Apart from the inversions centred at 0, this group also contains all positive homotheties with the same centre, and is isomorphic to the multiplicative group $\mathbb{R}^\star$ of non-zero real numbers.

151, 152. Straight lines that do not pass through the centre of the inversion, go into circles that pass through the centre, and vice versa.

153. This fact is proved in Chapter 7, Theorem 18 (see page 212).

154. The length of an orbit can be 2, 3, 6 or 12. See Figure 16 and the discussion below.

155, 156. Derive explicit formulas for the composition and the inverse transformation.

157. $(-3, 0)$.

159. (a) Yes, for example $y' = 0$. (b) No.

160. $y = 2x + C$, $y = \sin x + C$, $y = -1/x$.

161. A family of straight lines for equation (36) and a family of hyperbolas for equation (37).

162. Circles centred at the origin. The corresponding differential equation is $y' = -x/y$. It is defined everywhere except for the line $y = 0$. There is no such differential equation defined in all the plane.

163. $y = 1/(C - x)$.

164. $v = x + y$, $u = x$.

165. $y = -x^4/2 + x^2 \log x + Cx^2$.

166. $y = e^{x^2/4} \left(\int e^{-x^2/2} dx + C\right)^{1/2}$.

167. $y = \tan(ax + C)$.

168. $y = \dfrac{1}{x^2 \tan(1/x + C)} - \dfrac{1}{x}$.

169. Find n such that $-(n+3)/(n+4) = -4$.

170. Note that
$$\frac{1}{q(n)+2} = \frac{1}{n+2} - 1,$$
whence $\dfrac{1}{q^k(n)+2} = \dfrac{1}{n+2} - k.$

171. See Problem 67.

172. (a) $\dfrac{dr}{d\phi} = \dfrac{(yy'+x)\sqrt{x^2+y^2}}{xy'-y}.$

(b) The solution is given implicitly by
$$(x^2+y^2)(C - 2\arctan\frac{y}{x}) = 1.$$

173. $g_t \circ g_s = g_{t+s} = g_{s+t} = g_s \circ g_t.$

174. Yes. Assuming that the centre is 0, the correct parametrization is given by $g_t(x,y) = (e^t x, e^t y)$.

175. $x_t = x\cos t - y\sin t$, $y_t = x\sin t + y\cos t$.

176. These formulas define the group of spiral homotheties. Its orbits are logarithmic spirals.

177. The orbits are straight lines $x + y = \text{const}$. The group property follows from the Vieta theorem.

178. $y' = f(y/x)$, where f is an arbitrary function.

179. $y' = \dfrac{xf(\xi)+y}{x-yf(\xi)}$, where $\xi = \arctan\dfrac{y}{x} - \dfrac{b}{a}\ln\sqrt{x^2+y^2}$.

180. The function x^2+y^2 is a universal invariant of this group action.

181. The function ξ from the answer to Exercise 179.

182. $y = Cx^2 - x$.

183. Yes.

184. Use the group of homotheties with centre at the intersection point of the lines $ax+by+c=0$ and $a_1x+b_1y+c_1=0$.

185. Use the group $x_1 = e^{2t}x$, $y_1 = e^t y$. Answer:
$$y = \sqrt{Cx^2 - 2x}.$$

186. $y = \dfrac{x+C-1}{x+C}e^x.$

Index